Praise for
BIZARRE

"*Bizarre* is a collection of stories of how the brain can create zombies, cult members, extra limbs, instant musicians, and overnight accents, to name a few of the mind-scratching cases. After reading this book, you will walk away with a greater appreciation for this bizarre organ. If you are a fan of Oliver Sacks' books, you're certain to be a fan of Dingman's *Bizarre*."
— Allison M. Wilck, PhD, Researcher and Assistant Professor of Psychology, Eastern Mennonite University

"Through case studies of both exceptional people as well as those with disorders, *Bizarre* takes us on a fascinating journey in which we learn more about what is going on in our skull."
— William J. Ray, PhD, Emeritus Professor of Psychology, The Pennsylvania State University, author, *Introduction to Psychological Science, Research Methods for Psychological Science,* and *Abnormal Psychology*

"A unique combination of storytelling and scientific explanation that appeals to the brain novice, the trained neuroscientist, and everyone in between. Dingman explores some of the most fascinating and mysterious expressions of human behavior in a style that is case study, dramatic novel, and introductory textbook all rolled into one."
— Alison Kreisler, PhD, Neuroscience Instructor, California State University, San Marcos

"Dingman brings the history of neuroscience back to life and weaves in contemporary ideas seamlessly. Readers will come along for the ride of a really interesting read and accidentally learn some neuroscience along the way."

—Erin Kirschmann, PhD, Associate Professor of Psychology & Counseling, Immaculata University

"*Bizarre* is one of those rare reads that presents true neuroscientific information in a genuinely accessible manner. Dingman's entertaining writing style crossed with his depth of knowledge on the subject is sure to make this book a valuable asset to both the novice and the practiced scientist alike."

—Kate Anderson, PhD, Associate Professor of Psychology, Presbyterian College

BIZARRE

THE MOST PECULIAR CASES OF
HUMAN BEHAVIOR AND WHAT
THEY TELL US ABOUT HOW THE
BRAIN WORKS

Marc Dingman, PhD

NICHOLAS BREALEY
PUBLISHING

BOSTON • LONDON

First published in 2023 by Nicholas Brealey Publishing
An imprint of John Murray Press

An Hachette UK company

27 26 25 24 23   1 2 3 4 5 6 7 8 9 10

Text illustrations created by Marc Dingman

A CIP catalogue record for this title is available from the British Library

Library of Congress Control Number: 2022942592

ISBN 9781399801201
US eBook 9781399805353
UK eBook ISBN 9781399801225

Printed and bound in the United States of America.

John Murray Press policy is to use papers that are natural, renewable and recyclable
products and made from wood grown in sustainable forests. The logging and
manufacturing processes are expected to conform to the environmental regulations
of the country of origin.

John Murray Press Ltd
Carmelite House
50 Victoria Embankment
London EC4Y 0DZ
Tel: 020 3122 6000

www.nbuspublishing.com

For Michelle.
Without your love and support,
I truly don't know where I would be.

CONTENTS

INTRODUCTION

At about 11:40 a.m. on a hot, cloudless August day, 25-year-old Charles Whitman took the elevator to the top floor of the Main Building at the University of Texas (UT) at Austin. It was 1966, and at that time the Main Building, known simply as "the Tower" to students and locals, was the second tallest building in Austin—rising 307 feet into the Texas sky from the center of the UT campus.

Whitman was an Eagle Scout, ex-Marine, and a UT student—a six-foot-tall, muscular, blond-haired man who was generally well liked. He used his university ID to get past a security guard, who let him wheel an army-style footlocker on a dolly into the tower. Unbeknownst to the guard, the footlocker contained an arsenal of weapons.

Whitman reached the 27th floor, then took three steep half-flights of stairs up to the observation deck, which wraps around the 28th floor of the tower. When he got to the reception area of the observation deck, he was greeted by the receptionist, 51-year-old Edna Townsley. He immediately attacked her, fatally injuring her with a blow to the back of the head—probably delivered with the butt of his rifle. Minutes later, a group of visitors arrived to take in the views of the city available from the tower. Whitman fired at them with a sawed-off shotgun, killing two and critically injuring two others.

Charles Whitman as he appeared in the 1963 edition of the University of Texas student yearbook.

1

Then Whitman stepped out onto the observation deck, opened the footlocker, and spread his arsenal out on the floor. He had a collection of pistols and rifles and about 700 rounds of ammunition. Whitman selected a rifle that was designed for long-range accuracy. At 11:48 a.m., he began shooting at people walking on the UT campus hundreds of feet below.

His first shot ripped through the stomach of a pregnant woman, Claire Wilson, immediately killing her unborn son. As Claire fell to the ground, her boyfriend rushed to her to see what was wrong; he was shot in the back and died instantly. Whitman's next three victims were a physics professor, a Peace Corps trainee, and an undergraduate student.

That was all in the first 10 minutes of Whitman's terroristic onslaught. He continued his random attacks on the passersby below for over an hour and a half before police stormed the tower and shot him dead. When it was over, Whitman had killed 14 people (including Claire Wilson's unborn child) and injured more than 30 others. Another student, whose kidney had been severely damaged from one of Whitman's rounds, died in 2001; his death was ruled a homicide.

Of course, the first question on everyone's mind after such a tragedy is: Why? What would lead an architectural engineering student who many considered a "nice guy" to commit this heinous crime?

When police began investigating, more horrific details came to light. Whitman had also used a large hunting knife to kill his mother and his wife in the early morning hours on the day of the shooting.

While searching Whitman's home, police found a note Whitman had typed on a typewriter the night before the attack. In it, Whitman seemed to be trying to make sense of his homicidal urges. He wrote:

> I don't really understand myself these days. I am supposed
> to be an average reasonable and intelligent young man.

However, lately (I can't recall when it started) I have been a victim of many unusual and irrational thoughts. These thoughts constantly recur, and it requires a tremendous mental effort to concentrate on useful and progressive tasks...After my death I wish that an autopsy would be performed on me to see if there is any visible physical disorder. I have had some tremendous headaches in the past and have consumed two large bottles of Excedrin in the past three months.[1]

Whitman's wish for an autopsy was granted the day after his death. Although Whitman's case was already an intriguing one for sociologists and criminologists, the autopsy caused Whitman to be thrust to the forefront of debates on the brain and behavior. For, when doctors examined Whitman's brain, they found a large tumor impacting a structure called the *amygdala*, which plays an important role in emotional regulation. (We'll talk more about the amygdala in the chapters to come.)

This discovery prompted some to attribute Whitman's homicidal behavior to the presence of the brain tumor. Indeed, it seems plausible that Whitman's tumor could have influenced his amygdala in such a way as to cause unexpected personality changes, potentially leading to his vile actions.*

Others, however, are not so quick to blame Whitman's tumor for his crimes. Despite the accounts of Whitman being likable, he had a temper that sometimes frightened his wife, and he admitted to physically attacking her on two occasions. At the time of the shooting, he was also dangerously abusing amphetamine. It wasn't uncommon for him to stay awake for days taking large doses of the

---

* It's important to note that mass shootings are not all thought to be attributable to neurological problems or psychiatric disorders. In truth, mass homicide is a poorly understood act with complex influences that vary depending on the individual in question. Although brain dysfunction is frequently blamed for mass shootings, the evidence to support such assertions is often lacking.

drug, behavior that increases the likelihood of violent outbursts— and can even cause someone to lose touch with reality.

Regardless, Whitman's case is interesting from a neuroscientific perspective because neuroscientists know we cannot rule out the possibility that his killing spree was attributable to effects caused by his brain tumor. In fact, there are countless other cases throughout history where tumors, strokes, brain damage, and the like led to changes in personality that made the individual barely recognizable to those around them.

Probably the most famous of these is the case of Phineas Gage, a railroad foreman who in 1848 accidentally caused a small explosion that propelled a 3-foot-7-inch, 13-pound metal rod toward his head. The rod was tapered at one end, and it entered Gage's face below his left cheekbone, traveled through his skull, carved a hole in his brain, and exited from the top of his cranium with great force, landing some 75 feet away. Amazingly, Gage survived the accident and—after a period of several weeks where his prognosis was grim—eventually recovered almost all of his physical abilities except for sight in his left eye.

What happened next is disputed, as there are very few confirmed details of Gage's post-accident life (much of his later biography is based on hearsay). As the story goes, Gage's friends and family claimed the old Gage was lost forever after the accident. Previously responsible and conscientious, the post-injury Gage was reported to be impulsive, unscrupulous, and profane. He was unable to regain his position at the railroad due to his person-

Phineas Gage holding ality changes, and he spent the next 12 years *the metal rod that was* working odd jobs—including a stint at P.T. *propelled through his* Barnum's American Museum in New York, *skull and brain after* where he put himself on exhibit with the rod *an explosion in 1848.* that caused his injury. He eventually died in

1860 during a seizure that was likely related to the brain trauma he suffered in his accident.

The tale of Phineas Gage has become about as close to neuroscience mythology as you can get, with the details of his personality changes being embellished over the years to suit the intentions of whoever is telling the story. Nevertheless, Gage is frequently held up as an example of how the integrity of our brain determines who we are at the most fundamental level, and how disruptions to brain function can drastically alter the core elements of our personality.

Gage's and Whitman's cases are intriguing, but they're also surrounded by controversy because many of the details (about both their behavior and their brains) remain uncertain. In this book, however, we'll examine a number of lesser-known—but more precisely documented—cases of people who, because of some deleterious influence on their brain, had their typical experience with the world categorically transformed. But the resultant changes we'll explore involve much more than just personality. Instead, we'll focus on the strange—indeed, the downright most peculiar—consequences that can appear due to abnormalities in brain function. You'll meet patients who endure unwelcome additions to their mental life, such as the sense that their body has transformed into that of another species, the belief that they are no longer living, or the occurrence of hallucinations so vivid they put the strongest psychedelic drugs to shame. Others have lost critically important faculties, like the ability to recognize the faces of people they've known all their lives, the capacity to distinguish between a mirror and the real world, or the capability to form any type of image in their head.

While most of the unusual phenomena I'll discuss in this book are brought about by some adverse effect on the brain—such as trauma, a tumor, infection, stroke, or psychiatric condition—others are not the result of disorders at all. Instead, they are curious manifestations of an otherwise normal brain—the far end of the spectrum of human behavior. A few are even commonplace behaviors that all of us exhibit to some degree—often without awareness that we do, or at least without a good understanding of why. Indeed, you may be surprised to learn some of the weird things your brain does on a daily basis, seemingly without your cognizance or consent.

In fact, the only shared characteristic of all the behaviors I'll discuss in this book is that they are exceedingly strange, and the brain is held responsible for them. They represent what I consider to be the oddest collection of curiosities to emerge from the brain, and, if nothing else, they can be used as compelling evidence that the human brain is a powerful—but incredibly bizarre—organ.

If you don't hold this opinion already, then by the end of the book you likely will. In each chapter, I'll introduce a collection of extraordinary brain-related phenomena with a common theme. As examples, I'll use descriptions of real individuals—usually (but not always) patients with a particular medical condition—who exhibit some outlandish behavior. Often, I've fictionalized minor details of these cases. For example, I've frequently given anonymous patients names to make it easier to talk about them. (I've attempted to make these names appropriately suited to the region in which the case was documented, in the hope of accurately representing the patient's culture.) In a few instances, I've added trivial details or even a little dialogue to paint a clearer picture of what the patient was experiencing. But never have I exaggerated the specifics in such a way as to make them inconsistent with the case's true presentation. In other words, as implausible as some of these cases may seem, they are all representative of real behavior exhibited by real people.

It's worth stressing those last couple of words: *real people*. I wanted to write this book because the behavior described within it is fascinating from a neuroscientific—or even just a human— perspective. But it's easy to get wrapped up in the curious details and forget that some of the disorders I'll discuss cause considerable suffering. So, while I've tried to avoid writing about cases in an overly grave manner simply to make the book more entertaining to read, I want to emphasize the respect I have for the individuals who experience these conditions. I take their struggles very seriously. Far from sideshow attractions, many of the patients described herein are incredible examples of resiliency.

For each behavior I introduce, I will offer some explanation of what might be going on in the brain to cause it. I should point out, however, that most of the phenomena I'll discuss in this book are incredibly rare and/or poorly understood. Thus, the hypotheses I will propose to explain them are just that: hypotheses. They are not my own hypotheses; I have drawn them from the work of well-respected researchers. Still, for almost all of the aberrant behavior we'll explore in the pages to come, much more needs to be learned before we can confidently say what is happening in the brain to produce it.

Nevertheless, my hope is that this book will provide you with an interesting backdrop to learn a bit more about your brain. After all, it was the most unusual cases in neuroscience that drew me to the field. They fascinated me and created an insatiable desire to learn how such strangeness could possibly come out of this enigmatic organ in our heads. Thus, if you walk away from this book with an increased interest in neuroscience, I'll consider it a success. But perhaps you'll also develop a better understanding of how your brain works—and maybe even a greater appreciation for the stability of the reality you experience.

After all, many of the cases in this book illustrate that the reality we're so familiar with is also incredibly tenuous. We move through our lives maintaining a somewhat intentional unawareness of how one unexpected event can completely transform who we are and how we experience the world. Many of the neurological changes you'll read about in this book are the type that no one ever expects will happen to them; yet they do happen to people every day. And just like the cases I'll discuss, your mental life can be drastically and unpredictably altered in a matter of minutes, and you may never be the same.

1

IDENTIFICATION

In the late eighteenth century, a 70-year-old woman named Hilde was preparing a meal in her kitchen in Denmark when her brain suddenly found itself in the quite undesirable state of being deprived of blood. This, unfortunately for Hilde, is a condition human brain cells have pretty much no tolerance for. Without blood, neurons (the primary cells in our brain) quickly begin to run short of essential substances like oxygen and glucose; within minutes, they start to die. As the privation continues, neurons begin to expire at an unsettling pace—almost two million a minute. In that same minute, about seven-and-a-half-miles of neural fibers (the long extensions of neurons that carry signals from one cell to the next) can be wiped out.[1] In short, a lack of blood is devastating for the brain. The term for this grim situation is a stroke, and Hilde's sent her into a coma.

The details of Hilde's case come from a scientific paper published in 1788. The paper doesn't mention her family's reaction when Hilde awoke from her coma four days later, but it's safe to assume they were relieved. It's also likely, however, that their relief took a great blow when Hilde began to insist that she was dead. To be clear, Hilde was not claiming she had a near-death experience— that she had seen the tunnel and the light and at the last moment been yanked back into the land of the living—but that she, at the time she was talking to her family, was not alive.

We know about Hilde's case through the writings of an

eighteenth-century Swiss scientist named Charles Bonnet.[2] Bonnet was a lawyer by profession, but like most of the great minds of his day, he dabbled in a number of different fields, embarking on new scientific pursuits as casually as many of us take on watching a new television series. Surprisingly, he had a great deal of success going about things in this way.

Bonnet was, for example, the first to confirm that sex is not a prerequisite for procreation when he documented asexual reproduction in the aphid, a pesky bug well known to (and generally despised by) gardeners. In other entomological investigations, he made significant contributions to understanding how insects breathe. Then, he turned his focus to botany, where his work laid the foundation for the understanding that leaves are the sites where carbon dioxide and oxygen pass into and out of a plant. Not bad for someone with no formal scientific training, whose scientific exploits were something of a hobby.

Fortunately for our purposes, Bonnet also took an interest in unusual human cases such as Hilde's. Truth be told, her name was not Hilde. Or perhaps it was, as Bonnet never mentioned her name in his description of her. Like so many medical cases that have found their way into the scientific literature, Bonnet probably withheld Hilde's real name to protect her anonymity. But I've given her a common Danish name to make it easier to discuss her case.

Prior to Hilde's stroke, she had no serious mental health issues, which made her strange behavior all the more perplexing. Her family tried to convince her that she was not dead. After all, she was sitting up and talking. She had recovered; it should have been a time to celebrate life. But Hilde was having none of it. She became agitated and angrily chastised her family for not having the decency to give her a proper funeral. She demanded they dress her for burial, put her in a coffin, and arrange a send-off fitting for a woman of her stature.

Everyone hoped the delusion would fade with time, but Hilde's insistence only became more and more emphatic, soon turning into threats. It began to seem that conceding to her wishes might be the only way to calm her down.

So, her family reluctantly did just that. They wrapped her in a burial shroud (burial shrouds were apparently a thing in eighteenth-century Denmark) and acted as if they were planning to bury her. Hilde spent some time fastidiously rearranging her shroud, complained in a schoolmarmish way that it wasn't white enough, then lay peacefully until she drifted off to sleep.

Her family undressed her and put her back into bed, hoping the episode was finally over. But when Hilde awoke, she picked up right where she had left off, immediately insisting that she needed to be buried. Her family—not about to go so far as to put Hilde into the ground (even if it were a mock burial designed only to appease their quarrelsome patient)—decided the only thing left to do was wait for the strange delusion to pass.

Eventually, it did—but only temporarily. Every few months, Hilde would be convinced all over again that she was dead, puzzled that she was the only one who was able to recognize her demise.

The walking dead

Nothing quite like Hilde's case appears in scientific writings before Bonnet's report, but many comparable cases have been documented since. Indeed, we have seen enough similar cases to be confident that Hilde was not suffering from a one-of-a-kind neurological quirk, but from a disorder with distinctive and somewhat predictable symptoms. The condition is so rare that it's difficult to come up with a reliable estimate of how often it occurs,[3] yet it's common enough to have earned a name: *Cotard's syndrome*.

The eponym of the disorder is the French neurologist Jules Cotard, who lived in the second half of the nineteenth century. In 1874, Cotard was working in a town outside Paris when he encountered a patient who said she had no brain, nerves, or internal organs. She claimed she did not need to eat to survive and that she was immune to pain. The latter assertion seemed to have some validity to it: Cotard wrote that he could "deeply push pins" into her skin without evoking any reaction[4] (doctors really could get away with a lot more in the 1800s than they can now).

The patient, whom Cotard referred to as Mademoiselle X, did not believe she was dead, but instead suggested she was in some sort of limbo state—neither dead nor alive. She feared she would be stuck in that state of nonexistence forever and actually longed for true death, which she thought—without any very convincing evidence—she could only attain by being burned alive. She tried to prove this latter point on her own, but fortunately she was unsuccessful.

Intrigued by Mademoiselle X, Cotard searched for earlier descriptions of similar cases. To his surprise, he dug up several reports of patients who claimed they were rotting away, lacking blood or a body, damned to exist in a state of eternal oblivion, or experiencing some similar existential catastrophe. Cotard decided they were all suffering from a related condition. He called it *délire des negations*, or "delusions of negations." A delusion, of course, is a belief that is clearly false yet seems to be undeniably true to the patient in question, and Cotard used the word *negations* to refer to the most distinctive symptom in these patients: the denial of the existence of things that seem, to most of us, to be indispensable to life.

Several years after Cotard died, another scientist writing about delusions of negations referred to the condition as Cotard's syndrome. Since then, it has variously gone by Cotard's syndrome, *Cotard's delusion*, and sometimes even *walking corpse syndrome*. Scientists generally shun the last term, since (in addition to the unscientific hyperbole, which is the type of thing that makes most scientists cringe) claiming to be dead is just one of the many ways Cotard's syndrome can manifest in patients. Some of the other existential assertions mentioned above are actually more common.

Many other symptoms can occur in Cotard's syndrome as well, such as apathy, heightened or dulled senses, loss of hunger or thirst (and consequent self-starvation or dehydration), hallucinations, anxiety, severe depression, self-harm, and suicidal ideation. This is, unfortunately, an abridged list. Nevertheless, the denial of existence is what can really make descriptions of Cotard's syndrome patients sound like fiction.

Unusual complaints

One 28-year-old stockbroker, whom we'll call Will, was in a serious motorcycle accident in October 1989. He suffered brain trauma that sent him into a coma, and although he regained consciousness within a few days, he spent the next several months in the hospital recovering from the insults to his brain as well as battling multiple infections associated with his other injuries.

By January, however, Will had made an impressive recovery and was ready to be discharged from the hospital. He did have some lasting physical issues, such as difficulty moving his right leg and partial blindness. His most troubling complaint, however, involved his thoughts: he was quite certain he was dead.

In a desperate attempt to aid in Will's convalescence, his mother took him on a vacation to South Africa. But the South African heat caused Will to believe he was (literally) in Hell, further convincing him that he must be dead. When his mother incredulously asked him about his cause of death, Will offered up several possibilities. It seemed to him that a blood infection (which had been a risk early in his recovery) was plausible, as were complications linked to a vaccination he had received for yellow fever. He also suggested that he might have died from AIDS, even though there was no indication he had HIV or AIDS.

An unshakable feeling gripped Will—a feeling that everything around him was not, for lack of a better term, real. He had trouble recognizing people and places he had been familiar with before the accident, which contributed to his sense that he was living in a strange, foreign world. Even his mother didn't seem like herself. While he was in South Africa, in fact, Will concluded that she wasn't. He decided his mother was still asleep at home and that her spirit was accompanying him to show him around the underworld.[5]

Julia, a 46-year-old woman with severe bipolar disorder, entered the hospital convinced that her brain and internal organs had vanished. She felt she no longer existed, and all that remained of her was a body with nothing in it. Her "self" had disappeared, and thus she was (for all intents and purposes) dead. She was afraid to take a bath or shower because she thought her empty body might slip down the drain.[6]

Kevin was a 35-year-old man who had become increasingly depressed for several months before his thinking started to turn delusional. First, he suspected his family was organizing a secret plot against him. Then he decided he had died and gone to Hell but left his body behind. His body, he thought, was now an empty, bloodless shell. Determined to make a convincing demonstration of this, he grabbed a knife from his mother-in-law's kitchen and began stabbing himself repeatedly in the arm. At this point, his family wisely called an ambulance and had him hospitalized.[7]

Making the world make sense

Clearly, something in the brain of a Cotard's syndrome patient is not quite right. A serious neurological event (e.g., stroke, tumor, brain trauma) or psychiatric condition (e.g., depression, bipolar disorder, schizophrenia) often precedes the disorder. These types of problems, however, do not usually lead to Cotard's syndrome, and neuroscientists have yet to determine with certainty what it is that makes a Cotard's syndrome brain so different. The fact that Cotard's syndrome never looks quite the same from one patient to the next further complicates the issue. Nevertheless, a handful of common symptoms might provide clues to understanding the disorder.

Patients with Cotard's syndrome often complain that the world around them appears strangely unfamiliar. The people and places they know do not generate the spark of recognition most of us experience when encountering something we have encountered many times before. A Cotard's syndrome patient, for example, would likely recognize his mother's face, but something about her might seem foreign. Some intangible—yet essential—quality would be missing, causing the patient to lack the emotional response he would expect to feel upon seeing one of the most important people in his life.

Patients can also feel detached, as if they are an observer of the world but not a participant in it. The technical term for this is *depersonalization*. Additionally, everything may take on a surreal quality, causing patients to believe they are living in a life-like dreamscape—a symptom referred to as *derealization*. The unfamiliarity, depersonalization, and derealization Cotard's syndrome patients experience makes for a drastically altered reality. This, as you can imagine, is a lot for your brain to handle.

When your brain experiences such jarring symptoms, it scrambles to make sense of them. To your brain, having a rational interpretation of life's events is critically important. Without it, the world quickly becomes an unpredictable, incomprehensible, and ultimately unbearable place. Thus, your brain will seek out clear explanations for the things you experience with an almost desperate tenacity. When it has a difficult time finding a logical explanation for some element of your experience, it does the next best thing: it makes something up.

None of us are immune to this type of fabrication; we do it all the time without realizing it. For example, research suggests we make countless decisions daily—about everything from when to have a snack to who to go on a date with—without truly thinking much about them at all. It's almost as if we spend a great portion of our lives on autopilot. When asked after the fact about a decision we've made, however, our brain almost always comes up with a good explanation to justify our choice. Sometimes it's complete nonsense.

In one study, researchers showed male and female volunteers a pair of female faces and asked them to decide which face they found more attractive. Immediately after the study participants made their decision, researchers showed them the picture they had chosen and asked them to explain their choice. Unbeknownst to the participants, 20 percent of the time experimenters surreptitiously switched the pictures and asked the subjects to justify a selection they had not actually made.

Most participants didn't notice the chicanery. Instead of objecting, they usually just came up with some impromptu rationalization for the choice they assumed they had made, such as "she looks very hot in this picture," or "I thought she had more personality." (The pictures, by the way, were quite dissimilar, so the participants were not simply mistaking one face for another.[8])

This type of unintentional fabrication is known as *confabulation*, and your brain does it more often than you'd like to believe. While the reasons for confabulation can vary, it seems to be one strategy the brain uses for making sense out of events it lacks a clear explanation for. Neuroscientists believe something similar happens in Cotard's syndrome.

According to this perspective, Cotard's syndrome starts with brain dysfunction linked to one of the conditions mentioned earlier, such as trauma, a tumor, etc. This dysfunction causes symptoms of derealization and depersonalization, which make the patient feel as if everything around them is unfamiliar and lacking the quality of "realness" they expect. The patient's brain, trying to make sense of its experience, frantically searches for an explanation.

For unclear reasons, Cotard's syndrome patients tend to turn their focus inward, assuming that if there is something wrong with their experience, the problem likely originates with them. And then, for reasons that are even more uncertain, the explanation their brain lands on is that they are dead, decaying, possessed, or something else along those strange existential lines.

This hypothesized sequence of events may sound a little far-fetched. After all, a symptom like derealization is not that uncommon; many people (up to 75 percent of us, according to some estimates[9]) experience similar—albeit very transient—episodes at some point or another. But almost no one who has an episode of derealization ever thinks they are dead. Clearly, there must be something else going on in the brain of a Cotard's syndrome patient. Neuroscientists believe it might involve the failure of an important plausibility-checking mechanism.

An illogical brain

Although our brains may sometimes come up with faulty explanations for the events in our lives, typically we do not conjure up an interpretation that blatantly flies in the face of rationality. There seems to be a mechanism in our brain that evaluates our logic to make sure it passes the sniff test for plausibility.

For most people who experience symptoms such as derealization and/or depersonalization, that plausibility-checking mechanism would enable them to promptly dismiss the idea they were feeling disconnected because they were dead; the notion would be recognized as a ludicrous proposal and probably never thought of again. In patients with Cotard's syndrome, however, the plausibility-checking mechanism appears to malfunction. When their brain attributes their detached feelings to them being dead, the idea somehow retains its credibility, and the brain accepts the explanation as valid. The result is a belief that the rest of us consider unquestionably delusional.

Physicians looking for brain damage in patients with Cotard's syndrome (and—as we'll see—several other disorders characterized by outlandish delusions) often find it on the right side of the brain.

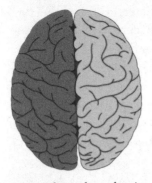

A brain from above showing both cerebral hemispheres. The left hemisphere has been darkened to make the division between the two hemispheres more distinct.

This has led neuroscientists to hypothesize that the right side of the brain is home to our plausibility-checking mechanism.

Our brain, you see, is divided into two halves known as *cerebral hemispheres*. The division is quite literal, as there is a large fissure that nearly slices the brain in half to create the partition. The two hemispheres of the brain appear to be identical at first glance. A trained neuroanatomist, however, can detect some asymmetries with the naked eye. Under a microscope, the discrepancies become even more distinct. Perhaps it's not surprising, then, that the two cerebral hemispheres also display some differences in function.

Awareness of these functional differences has long been fodder for inaccurate generalizations and exaggerations about dissimilarities between the right and left cerebral hemispheres. Take, for example, the assertion that some people use their right brain more (i.e., they are "right-brained") and so tend to think more creatively, while others are "left-brained" and likely to be more logical. While this idea is oft-repeated, neuroscientists consider it a myth. In truth, we do not typically display an overall bias in brain activation; we use both halves about equally overall.

There are, however, some functions (such as certain aspects of language) that depend more on one cerebral hemisphere than the other. Thus, the hypothesis that Cotard's syndrome is linked to right hemisphere damage is not improbable. The connection of Cotard's syndrome (and presumably, the brain's plausibility-checking mechanism) to the right hemisphere, however, remains hypothetical—an observation supported by many—but not all—cases of Cotard's syndrome that neuroscientists have studied in depth.

Regardless of where it is located, the hypothesized plausibility-checking mechanism plays an important role in a general model for what happens when patients develop Cotard's syndrome. First, some brain dysfunction leads to symptoms of detachment, such as derealization and depersonalization. The brain, as it is wont to do, tries to come up with explanations for what is going on. But the ability to scrutinize those explanations and discard the ones that do not align with rational thought is also impaired. Thus, the brain concocts the outlandish rationale that its body is dead (or possessed, or decaying, or whatever) and fails to reject it as invalid.

Some think this multistage process in the development of delusions could occur in certain other delusional disorders as well. The result is a collection of conditions with symptoms that rival those of Cotard's syndrome in their strangeness.

A world of imposters

Alex was 44 years old in early 1974 when his life began a downward spiral. He had recently been through a period of unemployment and was struggling financially, but it was when he found a job again that things really started to fall apart. His economic hardships had scarred him mentally, and he found himself obsessing about money constantly. He was in perpetual fear that he was on the brink of losing his new job—a preoccupation that caused him to be unable to sleep for more than two hours a night.

Clearly, Alex was suffering from some psychiatric issues, but things were about to get much worse. Amid all this psychological distress, Alex was hit by a car and sustained a serious head injury. As doctors performed surgery to stop the bleeding in his brain, they realized Alex would likely have lasting damage. A pool of blood in the right frontal part of Alex's brain had put increased pressure on sensitive brain tissue, killing brain cells in the process.

Alex remained hospitalized for 10 months after the trauma. But he improved substantially over that time, and doctors eventually gave him the OK to leave the hospital on weekends to visit his family at home. When the trips home began, Alex started to exhibit some peculiar behavior.

Upon returning to the hospital after his first visit with his family, Alex told his doctors that he lived in a different house than he had before the accident. This type of statement would not normally be a cause for concern, except for the problematic fact that Alex's family had not moved; the house he had visited was the same one he had lived in before his hospital stay. When asked to describe his new home, Alex explained that it was nearly identical to his previous one. He couldn't specify any clear differences, but he was certain it was, in fact, a different house.

Alex's physicians might have been less troubled by Alex's odd conclusions if he had focused his unusual logic solely on his house. But Alex also claimed that a different family occupied his new residence. Like the house, his second family was almost indistinguishable from the first. Alex asserted that the wives of both families had the same name, came from the same hometown, looked the same, and had similar mannerisms. He had five children in his new family; they were exactly like the children from his previous family, down to the birthmarks. Alex insisted, however, that he could tell the two families apart—although he could not explain how he was able to do that.

Surprisingly, Alex was generally unperturbed by the situation. He happily accepted his new family without any apparent reservations. He was uncertain why his first wife had left, but he was also thankful she had found someone else to take her place.

He was also aware of how unbelievable his claims sounded. Here is a real exchange he had with one of his doctors:

Doctor: Isn't that [two families] unusual?
Alex: It was unbelievable!
Doctor: How do you account for it?
Alex: I don't know. I try to understand it myself, and it was virtually impossible.

Doctor: What if I told you I don't believe it?

Alex: That's perfectly understandable. In fact, when I tell the story, I feel that I'm concocting a story... It's not quite right. Something is wrong.

Doctor: If someone told you the story, what would you think?

Alex: I would find it extremely hard to believe... [10]

And yet, despite this awareness, Alex maintained the belief. Several months later, when physicians interviewed him again, he was adamant that he had not seen his real family for quite some time. He claimed his second family had come to occupy the prominent familial role in his life.

Alex was suffering from *Capgras syndrome*, a disorder named for the French psychiatrist Joseph Capgras, who first described it in 1923. Capgras syndrome patients display a unique behavioral aberration: they believe that people close to them (such as a spouse, children, parents, or siblings) have been surreptitiously replaced by imposters who look and act just like their loved ones. The patients claim they can tell these imposters apart from the "real" people they have replaced, often by a trivial difference in appearance or behavior—or by some intangible quality the patient finds it impossible to explain.

Over time, Capgras syndrome patients typically begin to identify an increasing number of imposters in their life. In some cases, imposters overrun their world. The first Capgras syndrome patient ever described, whom Joseph Capgras referred to as Madame M, believed her daughter had been abducted and replaced by an imposter. Then, that imposter was replaced by another, who was also replaced repeatedly until Madame M encountered more than 2,000 imposters over a period of four years. Madame M also believed that her husband had been murdered and replaced by an imposter. She had been frustrated in seeking justice for her husband's death because the police force had been replaced by imposters as well.[11] The delusion can even extend to pets; one patient came to believe his poodle was an imposter.[12]

Individuals with Capgras syndrome tend to have relatively normal mental functioning otherwise. Their memory is typically intact, they can think clearly, and they can often even appreciate how absurd their delusion sounds (although this does not dissuade them from believing it). But there are other psychological qualities that are disturbed in Capgras syndrome.

Capgras syndrome patients often complain they do not feel an emotional connection to other people. Scientists have even confirmed this emotional numbness in studies that have found Capgras syndrome patients to lack a typical emotional response upon seeing someone they know.[13]

In other words, when you see a picture of your mother, there is something inside your brain that sparks up and generates emotional reactions such as *love, safety,* etc. (the emotions generated depend on the nature of your relationship with your mother, of course). Capgras syndrome patients, however, see a familiar face and experience little to no emotional reaction.

Thus, once again we have a disconnect between expectation and perception. The brain recognizes a familiar face but is also aware that seeing the face is not bringing about the emotional response it should. The brain rushes to explain this unexpected lack of emotion and arrives at the half-baked notion: "Well, if you don't feel an emotional connection to this person, then it must not be the person you think it is."

Typically, this type of explanation would be thrown out after some rational analysis, but it seems that Capgras syndrome— like Cotard's syndrome—involves a disruption in the brain's plausibility-checking mechanism. And, you guessed it, Capgras syndrome is also often linked to damage to the right hemisphere of the brain.[14]

Delusional misidentification syndromes

Researchers call Capgras syndrome a *delusional misidentification syndrome* because it combines a delusional belief with a clear deficit in identifying others, including those whose identities patients should be the most certain of. Cotard's syndrome is sometimes also included in the category of delusional misidentification syndromes because it involves perhaps the most egregious form of misidentification: misidentification of the self (as deceased, decaying, etc.).

There are other delusional misidentification disorders, which—not to be outdone—are bizarre in their own right. In the *Fregoli delusion*, for example, patients believe strangers are actually people they know—in disguise. In one case, a 66-year-old woman referred to as Mrs. C alleged that her cousin and his friend had moved into her neighborhood and begun stalking her. According to Mrs. C, her stalkers used disguises such as wigs, fake beards, and dark glasses to hide their identities as they secretively tracked her every move. Mrs. C was often late for her doctor's appointments because she had to take complicated routes to try to lose her pursuers.[15]

Patients with the *syndrome of subjective doubles* adopt the belief they have a double—as in an *Invasion of the Body Snatchers*–style duplicate—who looks just like them but leads a separate life. One hospitalized patient believed she had two doubles: one was being groomed to be the president of the United States while the other engaged in sadistic sexual acts in a different wing of the hospital in an attempt to tarnish the patient's reputation.[16]

Some even begin to accuse their own reflection in the mirror of shenanigans. Patients with a condition known as *mirrored-self misidentification* develop the belief that their reflection is a different individual altogether. They may suspect their reflection is spying on them and often become paranoid or fearful of their mirror image. One patient complained her reflection was stealing her clothes and jewelry,[17] and another (who believed his mirror image was a personification of his dead father-in-law) decided his reflection intended to harm him and his family. He often got into arguments with the mirrors in his house—until his daughters covered them all.[18]

Sometimes, delusional misidentification syndromes don't involve people at all. In cases of *delusional companion syndrome*, patients believe that certain inanimate objects are sentient beings with whom they converse and often form close relationships with. Frequently, the delusion focuses on stuffed animals or dolls. One 81-year-old woman, for example, began treating a teddy bear she was given as a retirement gift 17 years earlier as if it were alive. In talking to her doctor, she described the teddy bear as "a super person very interested in what is going on." Once, she took the stuffed animal out of the room when she met with her doctor in order "to maintain confidentiality." She repeatedly tried to feed the bear without success, but she did manage to get the stuffed animal to "absorb some fluid."[19]

Neurologically, one of the common themes across these various disorders is that patients tend to have suffered damage to the right hemisphere of the brain. Indeed, some neuroscientists have suggested that an impaired plausibility-checking mechanism might underlie all delusional misidentification syndromes. Although there seems to be merit to this hypothesis, there still is a lot to learn.

There is, for example, the question of precisely what parts of the right cerebral hemisphere are involved in plausibility-detection and how they work together to accomplish such a complex task. And, of course, there are the symptoms (such as derealization, emotional detachment, etc.) that leave the brain searching for answers in the first place; the neurological underpinnings of those symptoms are still not completely clear.

Cotard's syndrome and other delusional misidentification disorders are intriguing for their peculiarity if nothing else. But they, like many other disorders I'll discuss in this book, also act as a clear demonstration that our grip on reality is weaker than we would like to think. We take it for granted that our view of the world around us will be consistent, coherent, and rational. But we rely on properly functioning neural components to create that comprehensible worldview, and those components—like the parts to any machine—can fail. Thus, we are all only a head injury, stroke, or tumor away from being like one of the patients described in this chapter. What's more, the fragility of our conscious awareness extends far beyond cognitive functions like the accurate identification of someone you know. In the next chapter, we'll see how the perception of the shape and structure of your own body—and even the species it belongs to—can be distorted by your mind.

2

PHYSICALITY

When 24-year-old David checked into McLean Hospital— a large psychiatric institution situated just outside of Boston—he was convinced he was a cat. He was confident in this determination, in part, because his own cat Lola had told him so. Lola, you see, had taught David "cat language," enabling David to communicate with his feline brethren.

David was nothing if not committed to his feline persona. He acted like a cat, well, all the time—just as a cat would. He lived, hunted, played, and (regrettably) even engaged in sexual activity with cats.

He focused his romantic interests on cats as well; he had a crush on a tigress at the local zoo. Unfortunately, his love was unrequited, but David hoped one day to be able to rescue the tigress from her confinement and earn her affection in the process.

David's belief was not fleeting. He had been certain he was a cat for 13 years by the time he came to McLean Hospital in the 1980s. The doctors at McLean attempted to correct David's misconceptions with intensive therapy and an assortment of drugs. After six years of treatment, however, David's convictions remained as strong as ever.[1]

David was experiencing a condition known as *clinical lycanthropy*, a disorder in which one develops the delusion they have changed (or can change) into an animal. Traditionally, the word *lycanthropy* has been used to refer to the actual ability to transform into a wolf. Since ancient times, individuals in legends and mythology who possess this ability have been known as lycanthropes, a word that translates literally into "wolf men." Today, they're more commonly called werewolves.

Clinical lycanthropy, of course, refers to patients who only believe they can change into another creature (and not to real werewolves—who, depending on whom you ask, may or may not walk among us*). Despite the meaning of the word lycanthropy, the diagnosis of clinical lycanthropy is often used for patients who think they can transform into any animal (not just wolves). Some researchers, however, prefer to save the lycanthropy diagnosis for werewolf delusions, grouping others under the more general term clinical *zoanthropy*.

While clinical lycanthropy/zoanthropy is rare, more than 50 scientific reports have been published since the mid-nineteenth century describing patients who insisted they either had transformed or had the ability to transform into another animal. The type of animal their claims have focused on has really run the gamut. A search of the medical literature for clinical lycanthropy cases, for example, turns up descriptions of people who believed they could morph into a cat, dog, wolf, cow, horse, frog, bee, snake, wild boar, goose, bird, and even a gerbil.[2]

Lycanthropy through the ages

Clinical lycanthropy has a long history. Many consider Nebuchadnezzar II, the biblical king of Babylon who lived from around 642 BCE to 562 BCE** (the precise years of his life are uncertain), to be

---

* A 2021 internet survey of 1,000 Americans conducted by YouGov, a "research data and analytics group headquartered in London" (according to their website), found that 9 percent of Americans believe werewolves exist (the margin of error was approximately 4 percent).
** BCE stands for Before the Common Era. It is a nonreligious way of referring to the time before year zero—an alternative to BC (Before Christ).

one of the first recorded cases. The account of Nebuchadnezzar in the Bible states that the great warrior-king spent seven years acting like an animal and eating grass like a cow—apparently as a punishment from God for hubris. Many other cases of lycanthropy are described in classic and medieval sources, but the interpretation of these accounts is complicated by the tendency for people in those days to consider lycanthropy to be a real capability to transform into another animal (rather than a psychiatric disorder). Thus, the medical accuracy of older descriptions of lycanthropy leaves something to be desired.

Supernatural explanations for lycanthropy were the predominant view on the condition for millennia, but scientific interpretations began to prevail during the nineteenth century. Around this time, scientists started to view lycanthropy as a delusion and began recording cases as medical phenomena rather than folklore.

One of the first documented cases of clinical lycanthropy involved a man who was admitted to a psychiatric institution in France in the 1850s claiming he had transformed into a wolf. As proof, he pulled open his mouth to show that he had grown fangs. He also presented his body, which he asserted was covered in a lupine coat. His doctors made note of his claims but—not surprisingly—could not confirm them. All they saw was a slightly-hairier-than-average man who was disturbingly deranged.

The patient insisted on getting only raw meat for his meals, and when workers at the institution reluctantly obliged, he refused to eat it because it wasn't rancid enough. He became severely malnourished and eventually asked to be taken into the woods and shot like a dog. His doctors did not grant that wish, of course, but ultimately the patient died in the hospital due to undernourishment.[3]

As David's case demonstrates, however, lycanthropy is not just something of the distant past. People today are also occasionally stricken by the highly unusual belief that they are not people after all. Aleyna, for example, was 47 years old when her concerned family brought her to a hospital in Beirut in 2010. Aleyna's problems had started with an episode of depression that began shortly after her diabetic father had the toes of his right foot amputated. Aleyna was overcome with guilt about her father's amputation, even though she wasn't at all at fault. Excessive—and often unjustified—guilt is a common symptom of depression, and Aleyna's intense self-reproach coincided with the onset of her depressed mood.

Aleyna's doctor prescribed her an antidepressant, but after several weeks of taking the medication Aleyna hadn't shown signs of improvement. In fact, instead of getting better, she developed a strange and somewhat concerning mannerism: she would frequently stick out her tongue for no apparent reason and then quickly retract it back into her mouth. Soon after the appearance of this new habit, she told her family she had transformed into a snake. More specifically, she proclaimed that Aleyna was dead, and the devil had replaced her with a snake. (You might recognize after reading Chapter 1 that Aleyna was displaying symptoms of Cotard's syndrome along with her clinical lycanthropy.)

Aleyna refused to continue taking her antidepressant medications because they were "Aleyna's medications," and, needless to say, Aleyna was no longer living. Her family relied on their religious instincts and took Aleyna to a priest, who decided this was a textbook case of demonic possession. After a failed exorcism attempt, however, Aleyna's family brought her to the hospital.

Upon admission, Aleyna reiterated that she was a snake and said she was tempted to bite (and kill) members of the hospital staff. Indeed, she did try to bite the hands of several hospital workers during her stay. Doctors diagnosed Aleyna with depression with psychotic features and gave her medication typically prescribed to schizophrenic patients. Fortunately, these drugs seemed to do the trick: Aleyna recovered within several days and was released from the hospital, no longer experiencing any lycanthropic delusions.[4]

In another modern-day case, a 32-year-old Iranian man we'll call Amir entered the hospital claiming to be a dog. When speaking with doctors, Amir stated matter-of-factly that his wife had been turned into a dog as well. Along with Amir's alleged physical transformation, he said he had developed the exquisite olfactory abilities of a dog, and his newly refined sense of smell led him to notice that his two daughters' urine smelled like sheep urine. (Why Amir was so familiar with the smell of sheep urine—and why he was smelling his daughters' urine—is troublingly unclear.) Thus, Amir determined that his daughters had both transformed into sheep.

Like Aleyna, Amir also displayed symptoms of Cotard's syndrome, as he believed that his human body had died and been replaced by that of a dog. Doctors diagnosed him with bipolar disorder, Cotard's syndrome, and a rare variant of clinical lycanthropy—rare because in typical cases of clinical lycanthropy, delusions about being transformed into an animal don't usually involve thoughts that others have experienced a similar metamorphosis. After about two weeks of treatment in the hospital, Amir's symptoms began to fade. At a follow-up appointment two months later, he was in complete remission.[5]

The lycanthropic brain

Aleyna and Amir were both diagnosed with another psychiatric disorder along with clinical lycanthropy, which is par for the course. Conditions such as schizophrenia, depression, and bipolar disorder commonly precede the development of clinical lycanthropy. Depersonalization (the sense that one is a detached observer of the world around them, which we discussed as a key symptom in Cotard's syndrome) is also a common feature of clinical lycanthropy; some even consider clinical lycanthropy to be an extreme form of depersonalization.[6] Not surprisingly, clinical lycanthropy is also considered a delusion.

When it comes to the neuroscience of clinical lycanthropy, however, we are forced to rely mostly on conjecture. To this point, there have not been any studies that have specifically looked at what parts of the brain might be dysfunctional in a clinical lycanthropy patient. Given the delusional nature of the condition, we can speculate there is some degree of failure in the plausibility-checking mechanism discussed in Chapter 1. Otherwise, we would assume the belief that one has turned into a wolf, pig, snake, etc., would be discarded as a clearly irrational idea.

But according to neuroscientists, there is another brain mechanism that might be disrupted to cause clinical lycanthropy. That mechanism involves the creation of a mental representation of the body—something scientists often refer to as the *body schema*. Your brain uses this virtual likeness of your body to do things such as keep track of where your body is in space and maintain awareness of your body positions—processes that are going on in the background all the time without you noticing.

You can get a sense of what the body schema does by closing your eyes and moving your arm around slowly. Although you can't see your arm, you should have a good idea of where it is and what it is doing—you may even have a pretty clear image of what it looks like as it's doing these things. This is your body schema at work. The bodily awareness it creates makes it easier for you to move in a coordinated way, be cognizant of your posture, and interact with the environment.

An important characteristic of the body schema is that it is highly modifiable. This makes sense, as our bodies are malleable too; they change with age, activity, injury, and so on. Correspondingly, our body schema is frequently revised throughout the lifespan—due, for example, to typical experiences such as growth, as well as to trauma such as the loss of a limb.

But researchers believe neurological disorders can interfere with the body schema, and at least some cases of clinical lycanthropy might be linked to disruptions in the brain's ability to generate an accurate body schema. Indeed, some clinical lycanthropy patients claim they can *feel* their body changing, and that perception becomes part of the justification for their unusual beliefs.

One 21-year-old man, for example, sought medical help when he sensed that his chest had swelled and broadened, and his ribs had transformed into those of a dog. These perceived physical alterations were accompanied by changes in his behavior. By the time he met with doctors, he was growling at physicians and exploring his surroundings primarily through sniffing.[7] Another patient who believed he was a werewolf described his transformations (which typically occurred when he was emotionally upset) in this way: "I feel as if hairs are growing all over my body, as if my teeth are getting longer...I feel as if my skin is no longer mine."[8] In both of these cases, patients claimed they could feel their bodies changing, suggesting that a hallucinogenic disruption to the body schema might underlie the sensations that prompted their lycanthropic beliefs.

Phantom limbs

While clinical lycanthropy may involve a pathological changing of the body schema, sometimes the discordance between the physical body and the body schema can happen the other way around: the body changes, but the body schema doesn't properly account for those modifications. Judy, for instance, was 42 years old when she had her right arm amputated after a bad car accident. Several days following the amputation, she developed the enduring sense that her missing arm was still hanging by her side in a slightly bent position. She couldn't move the arm, but she could feel it—even though she knew her arm had been surgically removed.

Judy's complaint is surprisingly common; almost all amputees have the lingering sensation that their lost body part is still there.[9] The condition is known as *phantom limb*, but it doesn't only happen with limbs. It can occur after the loss of pretty much any body part, including fingers, eyes, breasts, genitals, and even teeth.[10]

Although having a phantom limb sounds like an interesting novelty, it becomes agonizing for most patients when they experience painful sensations (e.g., shooting pains, burning, cramps, aches) coming from their phantom limb. Sometimes these pains emerge spontaneously, while in other cases patients complain they experience pain because their phantom limb is stuck in an uncomfortable position and—since it no longer exists—they can't move it back to a comfortable one again. One soldier, for example, lost his hand when a grenade exploded in it. Afterward, he developed the sensation of a phantom hand stuck in a fist clenched so tightly that it was a source of constant pain.[11]

For someone like me who already must fidget constantly to get himself comfortable, this sounds like a nightmare, and for phantom limb patients, it truly can be. People who experience phantom limb pain have more anxiety, a lower quality of life, and a higher risk of depression.[12]

It's not fully understood how and why phantom limb occurs, but the body schema is at the center of one popular hypothesis. According to this view, when a limb is lost suddenly, the mental representation of your body remains intact. In other words, the body schema doesn't get updated to account for the lost limb. This creates an expectation—or even the distinct perception—that the limb is still present.

Multiple hypotheses have been proposed to explain why phantom limb so often results in pain, but there is no clear consensus. Pain might occur when (as mentioned above) the phantom limb gets stuck in an uncomfortable position the patient can't move it from. But it also might be due to faulty rewiring of the brain as it attempts to account for a lack of sensory input from the now missing limb, abnormal signaling generated from damaged neurons at the site of the amputation, or other factors altogether. Clearly, we have much more to learn.

The body in the brain

If it is true that there is a virtual representation of the body found within the brain, then a logical question would be: where in the brain is it? As with so many questions in neuroscience, the answer is: it's complicated. Like other sophisticated cognitive functions, the generation of a body schema doesn't seem to be something we can link to just one brain region. Instead, it's likely the result of multiple parts of the brain working together in an elaborate network.

This model of network communication is a good example of how modern neuroscience tends to look at the way the brain works in general. In the not-so-distant past, when neuroscientists had a behavior they wanted to explain in terms of brain activity, they strove to identify *the* area of the brain responsible for that behavior. Today, however, neuroscientists recognize that there are not many (if any) complex functions executed by one circumscribed area of the brain alone. Thus, when attempting to explain some aspect of how the brain works, we typically look for a collection of regions that collaborate to accomplish a task.

Of course, in accepting that the brain tends to operate through the activation of intricate networks, neuroscientists have also had to acknowledge that most brain functions are very complicated to untangle. Determining the structure of the network that generates the body schema has been an elusive goal of neuroscientists for decades.

Nevertheless, researchers have made some headway and identified several brain areas that may work together to formulate the body schema. The most recognizable of these regions is a large section of the brain called the *parietal cortex*. The term *cortex* (which is Latin for *bark* or *shell*) refers to the outermost layer of a bodily structure, but in this case we're referring specifically to the *cerebral cortex*, or the outer layer of the brain. At its thickest, the cortex only extends about four and a half millimeters deep into the brain from the surface of the organ. Even so, the cerebral cortex is far more than just the outer covering that its name implies. Neurons in the cerebral cortex are responsible for a catalog of functions ranging from sensory perception to some of our most complex cognition. The folding of the tissue in the cortex also creates the brain's most distinctive overt feature: the valleys and ridges that riddle the brain's surface, giving it a convoluted appearance.

The parietal cortex is situated toward the middle-rear of the brain;

the term is used inter-changeably with *parietal lobe*. Neuroscientists use the word *lobe* to refer to one of several discrete sections we divide the cortex into: the frontal lobe, parietal lobe, temporal lobe, and occipital lobe.* These divisions were originally made on a purely anatomical basis, but further research has taught us there are functional

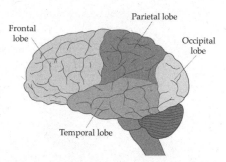

The lobes of the brain. Note that the term cortex *can be substituted for* lobe *in the labeled regions above. Thus, parietal lobe is synonymous with parietal cortex, frontal lobe with frontal cortex, and so on.*

differences between the lobes as well (not a surprise, given that the lobes all cover considerable portions of brain real estate).

The parietal cortex likely contributes to the formation of the body schema in multiple ways. First, it contains a region known as the *primary somatosensory cortex*—a key area for analyzing sensory information from the body. The primary somatosensory cortex, receives information about tactile sensations—the sensations of touch, pressure, vibration, etc., that you pick up through your skin. When you touch something, such as the surface of a table, the tactile data gets sent first to the primary somatosensory cortex, where information about the experience (e.g., What is the texture of the table? How hard is it?) is processed and made available to the brain.

The primary somatosensory cortex, then, is critical for enabling

---

* Sometimes a fifth lobe and sixth lobe, the *limbic lobe* and *insular lobe*, are included. I haven't introduced them here because they are not as common a designation as the four lobes I've mentioned, nor is a discussion of them necessary to comprehend any of the material covered in this book. Thus, I've opted for the more common and simpler approach of dividing the cortex into four lobes.

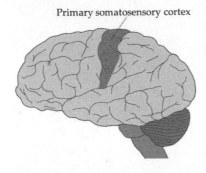

Primary somatosensory cortex

us to interact with the physical world. It also, however, gets information about a lesser-known—but essential—sense known as *proprioception*, which makes us aware of the body's position and where it is in space. Proprioceptive information is important for enabling you to move in a coordinated fashion and successfully navigate your environment. It's also crucial for the maintenance of the body schema; for your internal representation of the body to be accurate, it must be constantly updated using information about the position of your physical body.

Additionally, after the primary somatosensory cortex receives data about touch and proprioception, it communicates with nearby areas that get information from other sense modalities, such as hearing and vision. Your brain integrates all this sensory data to create a detailed awareness of your body, what it is doing, and the environment in which it is doing those things.

Strange denials

This idea that the parietal cortex plays an important role in the construction of the body schema is supported by the fact that damage to the region can lead to unusual disorders that involve disruptions in how the body is perceived. One example occurs as part of a condition known as *hemispatial neglect*, which is a common consequence of a stroke that affects the parietal lobe.

Patients with hemispatial neglect develop a curious inability to recognize the existence of things on one side of their field of vision. Even though they're still receiving information from their full visual field, they treat half of the world as if it isn't there. They might, for example, eat from only one side of their plate, put on only one shoe (leaving the other foot shoeless), or shave or apply makeup to only one side of their face. Often, this is all done without awareness of their oversight. In fact, sometimes these patients display a surprising degree of obliviousness that there is anything wrong at all.

It seems difficult to imagine someone being unaware they have this type of deficit. However, lack of insight into one's own health condition is common not only in hemispatial neglect, but in a number of other disorders as well. There's even a term for it: *anosognosia*, and it's one of the brain's more unusual tendencies.

Anosognosia translates into "without knowledge of disease." It occurs when a patient who is clearly experiencing some health condition is ignorant to their situation. Patients with anosognosia often display more than a lack of awareness; they might adamantly deny that anything is wrong, even if doing so requires some ludicrous rationalizations.

Some hemispatial neglect patients develop a specific type of anosognosia known as *anosognosia for hemiplegia*. Hemiplegia refers to paralysis of one half of the body, and in anosognosia for hemiplegia, patients are paralyzed on one side—but unaware of their state of paralysis. Despite clear evidence to the contrary, these patients will insist their body is fully functional. When asked to perform an action with their paralyzed limb, they might come up with some excuse, such as "I'm feeling very tired right now," or they might get agitated and try to change the subject.

In other cases, they develop delusional beliefs about why their body isn't working the way it should. They might, for instance, claim they're unable to move part of their body because it doesn't actually belong to them—a delusion known as *somatoparaphrenia*. Consider this real interaction between a doctor and a patient experiencing somatoparaphrenia associated with paralysis of the left side of their body:

Doctor: What about your left leg?

Patient: It was very difficult to begin with … to live with a foot that isn't yours.

Doctor: Why do you say that the foot is not yours?

Patient: I came to the conclusion that it was a cow's foot. And in fact I decided that they sewed it on. It looked and felt like a cow's foot, it was so heavy. But I adopted it. I'll take you home, I said.[13]

There are many examples in the scientific literature of patients with somatoparaphrenia coming up with fantastical narratives about their paralyzed limb. One patient said her hand was really her mother-in-law's hand. Another claimed a hand "was left on the subway" and ended up being surgically attached to her. Yet another insisted that gangsters had chopped off his brother's arm and threw it in the river before the patient discovered it lying next to him (he was at a loss as to how the arm had found its way out of the river and into his bed).[14]

Anosognosia for hemiplegia and somatoparaphrenia are both linked to damage to the parietal cortex and/or networks it acts as a key node in.[15, 16] Clearly, however, for patients whose anosognosia becomes delusional (such as patients with somatoparaphrenia), there must be some other mechanism involved for them to make the leap from lack of awareness to believing an outlandish explanation for that lack of insight. Some have hypothesized there must be a deficit in the type of plausibility-checking mechanism we discussed in Chapter 1—a deficit that could also be created due to the damage caused by an event such as a stroke.[17]

Based on the symptoms that occur when the body schema is disrupted, it seems to be an essential component of a healthy brain. But as we've seen, in some cases the body schema created by the brain is incongruous with the body. In phantom limb, the body schema remains complete despite the fact that a part of the body has been lost. And yet, other patients experience the opposite sensation: their body is intact, but their body schema is missing a limb. This can create a distressing—and often overwhelming—feeling that one's body has, for lack of a better term, an excess of limbs. This perceived mismatch can turn into an obsessive desire to change the body so it aligns with the body schema—no matter the cost.

A *bizarre obsession*

The Psychohormonal Research Unit (PRU) at Johns Hopkins University was established in the middle of the twentieth century to focus on what were considered sensitive topics at the time, such as gender identity and sex reassignment surgery. As one of the only medical facilities open to working with patients with sexual issues, the PRU was frequently contacted by people with problems not commonly seen in a typical hospital. Even for the staff in the PRU, however, Isaac's telephone call came as a bit of a shock.

Isaac phoned the PRU in the 1970s asking for a referral to a surgeon who would agree to amputate Isaac's left leg. As he stated in a subsequent letter, "Since my 13th year, my conscious life has been absorbed, with varying intensity, in a bizarre and...obsessive wish, need, desire to have my leg amputated above the knee."[18]

Isaac's interest in amputation had sexual overtones, which was why he contacted the PRU. He found the idea of amputees being able to navigate their daily lives sexually arousing. He was turned on by amputees walking on crutches. He sought out sexual encounters with amputees and masturbated to pictures of them as well.

But Isaac was not satisfied with just fantasizing about amputation or being sexually involved with amputees; he was intent on becoming an amputee himself. He searched desperately for a doctor willing to perform an amputation for him. The PRU, however, denied his request for a referral to such a doctor, and when Isaac started to lose hope that he would find a willing surgeon, he took matters into his own hands. What he did next is not for the squeamish, so if you count yourself among them you might want to skip the following two paragraphs.

Isaac got a piece of sharpened stainless steel, inserted it into his leg, and then used a hammer to drive it into his shinbone. He removed the piece of steel, leaving behind a cavity that extended from his skin to his bone. Isaac admired his work so far, but he was not finished yet.

Next, he mixed some pus from facial acne and nasal mucus together and inserted the mixture into the newly formed wound. Then he waited, hoping his efforts would lead to the type of severe infection that makes amputation unavoidable. Eventually, the infection he was yearning for appeared, and he eagerly proceeded to the local hospital, hopeful he would leave with one less leg. To his disappointment, however, the infection cleared while he was in the hospital, and he went home just as bipedal as before.

Desperately seeking amputation

In the late 1970s, researchers coined a term to describe the condition Isaac suffered from: *apotemnophilia*, a Greek word that roughly translates to "love for amputation." Since then, *Body Integrity Identity Disorder*, or BIID, has become the preferred way of describing the affliction. Initially, BIID was thought to be a sexual fetish; the first written descriptions of the disorder appear in letters to the magazine *Penthouse* in 1972 from readers with an erotic obsession with limb amputation. But many neuroscientists now believe BIID is a disorder that develops when someone has trouble incorporating one of their limbs into their body schema, causing them to feel an overpowering sense that the limb does not belong on their body. Interestingly, research suggests the body schema discrepancies in BIID may be linked to abnormal activity in the parietal cortex.[19]

Thus, BIID seems to be more of a neurological disorder than

a sexual fetish, but either way it's a condition that has the potential to completely alter someone's life. Not only do BIID patients become obsessed with losing a limb, but some of them actually succeed in their efforts. In the late 1990s, for instance, a surgeon in Scotland amputated the legs of two men simply because the men desired desperately to be amputees. The surgeon reported that the patients were extremely distressed before the operation, but afterward were happy with their decision and more content in their lives.[20]

It's rare, however, for patients with BIID to find a physician who is willing to help them satisfy their desire for an amputation. Without the aid of a medical professional, BIID patients are left to their own devices. Sometimes the outcome can be disastrous.

Carl, a 51-year-old civil servant in the UK, had been obsessed with a persistent desire to have a limb amputated since early adolescence. Carl's case is outlined in *The Journal of Hand Surgery*, complete with pictures that are more gruesome than what I'm used to seeing in medical journals.[21] When Carl was in his early 40s, he experienced a minor injury to his lower leg. Seeing this as an opportunity, he intentionally infected the wound; as a result, he had to have his leg amputated above the knee. Carl, however, was not satisfied—he was intent on an upper limb amputation as well. Again, a caveat: the next couple of paragraphs are not for those who are easily disturbed by gore.

To achieve his upper limb amputation, Carl took matters into his own hands (pun intended). First, he cut off his right pinky finger. That seemed to appease him for a time, because he didn't take any further action for about five more years. But then he mutilated the little finger of his left hand so badly that doctors had to amputate it. A couple years after that, he intentionally cut off his left ring finger. These finger amputations, as brutal as they might seem, did not realize Carl's true desire: to amputate his whole hand.

Eventually, Carl tired of this piecemeal approach, so he took an axe and chopped off his entire left hand. Concerned that doctors would simply attempt to reattach the hand, Carl then mutilated the severed limb with the axe, ensuring it was completely useless. He wrapped an elastic stocking around his arm as a tourniquet, then went to the hospital and asked for surgery to make it possible for him to have a prosthetic limb fitted to the stump. Doctors had little choice but to agree, as at that point there was no hope of saving Carl's hand. After the surgery, Carl claimed to be satisfied with the amputation. He was looking forward to the fitting of his new prosthetic limb.

Some BIID patients are content after finally obtaining the amputation they've been pining for. Others (like Carl after his leg amputation) experience a transient sense of gratification but have a recurrence of the desire. Another BIID patient, for instance, achieved the amputation he wanted by packing both his legs in dry ice for seven hours until they were severely frostbitten and hopelessly irreparable. Afterward, he ironically claimed to feel like a "complete person" for the first time in his life. Three years later, he was still happy with his decision; however, he had met a four-limb amputee, and—in an unlikely illustration of "the grass is always greener" idiom—the friendship sparked a new desire in him to amputate his left arm. With psychiatric treatment, his desire has abated a bit, but it hasn't disappeared completely.[22]

As strange as some of the disorders in this chapter may seem, they begin to make sense when considered in the context of the body schema. Indeed, the existence of an internal representation of the body seems logical and practical—an ingenious mechanism to aid our brains in navigating the world around us.

Nevertheless, the body schema is a difficult concept for me to wrap my head around. My mental perception of my body feels inextricably tied to my physical body in such a convincing way that I have a hard time thinking of them as anything other than one and the same. The research, however, suggests this is a common misconception, and that distortions in the neurobiological repre-sentation of the body can cause distortions in how the body feels to us. The neuroscience of our body representation, then, is another bit of evidence that suggests maybe nothing is quite as real as it seems.

3

OBSESSIONS

E lif was kneeling on the cold tile floor of an exam room in the hospital, crouched over a metal wastebasket, retching. This was the third time she had to interrupt her physical exam to vomit.

She had arrived at the hospital looking disheveled and pale, experiencing severe abdominal pain. When the doctor walked into the exam room, Elif was curled up in the fetal position, clutching her stomach, praying to fend off the paroxysms of nausea that coursed through her body.

Her condition seemed dire, but Elif's vital signs and initial tests all came back normal. Her pain, however, only got worse with time. Elif's doctor ordered a CT scan (a procedure that uses x-rays to acquire internal images of the body) of Elif's abdomen. That's when the description of Elif's case takes an unexpected turn.

The CT scan didn't reveal anything the doctor expected to find. She was looking for a tumor, intestinal obstruction, or severe inflammation—common underlying causes for the type of symptoms Elif was experiencing. But although the CT results didn't indicate one of these anticipated issues, they weren't normal, either. Elif's CT scan showed a number of highly unusual dark patches in Elif's stomach and small intestine. It appeared as if Elif had eaten something that had stuck to the lining of her gastrointestinal tract—not a common occurrence for typical foods.

Elif's doctor was puzzled and began asking Elif about her recent eating habits. Had she eaten anything out of the ordinary in the last few days? Could she have ingested something inedible by mistake? During this line of questioning, Elif's responses became a bit evasive. She appeared to be uncomfortable answering questions about her diet, and the doctor suspected she was hiding something.

Thinking she might be getting at the root of the problem, Elif's doctor pressed Elif about her dietary practices. Eventually, Elif made a startling admission: she had a penchant for eating cigarette ashes.

On a typical day, Elif confessed, she would eat the ashes from about two to three cigarettes. Her admission to the hospital, however, had been preceded by a particularly indulgent day when she had consumed the ashes of around 10 cigarettes. Her doctor concluded that the dark patches in her gastrointestinal tract represented areas of accumulated cigarette ash, which were also causing the symptoms that had brought Elif to the hospital. Elif's doctor recommended she meet with a psychiatrist, but Elif refused and left the hospital—without any indication that she planned to quit her dangerous habit.[1]

Elif had a condition called *pica*—a disorder characterized by a persistent desire to consume things that are typically considered inedible. The appetitive focus of a pica patient can vary drastically from case to case, ranging from something as harmless as ice to extremely dangerous fare like needles. In between, there are myriad substances pica patients have consumed that most of us would find either bland and generally unappealing—or downright repulsive.

On the more innocuous end of the spectrum, one patient abruptly developed a compulsive desire to eat raw potatoes and began incorporating three to five of them a day into her diet. She preferred them chilled (although room temperature potatoes would do in a pinch), and when going on trips would pack her own potatoes in a thermos full of ice water.[2]

While most of us would certainly consider this type of behavior to be unusual—imagine sitting down in your break room at work and watching a coworker unpack a few chilled potatoes from a thermos before biting into them raw—a potato obsession might not be all that shocking given that potatoes are part of a typical human diet. Consider then, the case of a 29-year-old pregnant woman named Charlotte (pica is more commonly seen in pregnant women than most other groups, for reasons we'll discuss shortly) who began eating burnt matchsticks during her third trimester. Unlike eating raw potatoes, this was not a harmless habit. Most matches contain substances that are dangerous to ingest, such as potassium chlorate, which is toxic in high doses.*

Doctors told Charlotte she was putting her baby's health at risk by continuing to eat the burnt matches, but Charlotte was unable to stop herself. Her cravings were overpowering; she couldn't suppress them no matter how hard she tried. After repeated failed attempts to change the behavior, Charlotte agreed to be induced once her baby reached full term to minimize the baby's exposure to the dangerous substances Charlotte was ingesting. The baby was born jaundiced and with some other minor complications but fortunately recovered well after treatment.[3]

Other patients compulsively eat their own hair. While this might sound relatively benign, when taken to an extreme it can cause serious problems. Hair is made up of a tough protein called keratin, which makes it resistant to the dissolving capabilities of your stomach acid. Thus, instead of getting broken down in the stomach, hair tends to accumulate in the gastric folds—extensions of overlapping stomach tissue that line the walls of the organ. Here,

---

* Ingestion of potassium chlorate has caused death in the past. One case from the early 1900s, for example, involved a man who ate an entire tube of toothpaste (which used to commonly contain potassium chlorate) on an empty stomach, in the process consuming enough potassium chlorate to bring about his untimely demise. The reasons for his behavior are unclear beyond the description of the man as a "mentally diseased army officer." (See S.A. Ansbacher, "A Case of Poisoning by Potassium Chlorate," *Journal of the American Medical Association* 96, no. 20 (1931):1681.) Fortunately, since the time of this case of poisoning, safer substances have replaced potassium chlorate in toothpaste and most other household items.

the hair mixes with mucus and food and shapes itself into a mucky clump. Over time (assuming hair is still being eaten), more hair will stick to the growing cluster. The resultant collection of hair is known as a *trichobezoar* by the medical community, but most of us would simply call it a hairball. In cases of compulsive hair-eating, patients can develop super-sized hairballs that get so large they interfere with gastrointestinal function.

One 7-year-old girl, for instance, came to the doctor with complaints of persistent abdominal pain and loose stools. Magnetic resonance imaging (MRI)—another method used to obtain detailed images of the internal body or brain—showed a large mass in the stomach, and doctors decided surgery was necessary to remove what they presumed was a stomach tumor. When the surgeon opened up the girl's belly, however, they found the mass was not a tumor but a tightly packed trichobezoar. The hairball filled the whole interior of the stomach and had a long tail that extended well into the small intestine. From end to end, it was 82 centimeters (over two and a half feet) long. It weighed 795 grams, or about 1.75 pounds.[4]

Potatoes, burnt matches, and hair are just a sample of the many strange items pica patients have been known to consume. Others include cotton balls, dirt, balloons, soap, toilet paper, twigs, clay, pebbles, mothballs, glass, feces, and toilet cakes (the little air fresheners often found in urinals).[5]

A trichobezoar (aka hairball) taken from the stomach of a 5-year-old girl who came to the emergency room with abdominal pain and vomiting (not the same case discussed in the text). The hairball formed a perfect cast of the stomach and extended into the small intestine. Adapted with permission. V. Gonuguntla and D. Joshi, "Rapunzel Syndrome: A comprehensive review of an unusual case of trichobezoar." Clin Med Res 7, no. 3 (2009):99–102. Copyright 2009 Marshfield Clinic. All rights reserved.

Although some of these examples are particularly unusual even for pica patients, pica itself is not as uncommon as you might expect, at least among certain groups. Studies have found, for example, that 20 to 30 percent of children under the age of six have practiced pica,[6] a statistic that is perhaps less surprising given the propensity of young children to put almost anything in their mouths. Pica prevalence does decrease with age, but it still can be relatively high in pregnant women and the intellectually disabled; according to some estimates, pica can be seen in over a quarter of all pregnant women[7] and intellectually disabled adults.[8]

The puzzle of pica

The big question facing scientists who study pica is: Why? What on earth would possess someone to eschew their bacon and eggs and turn to mothballs for breakfast? Perhaps, some researchers suggest, patients with pica develop their unusual urges out of an instinctual, but misguided, attempt to quell a nutritional insufficiency. Pica is more likely to occur along with iron deficiency anemia, so one hypothesis is that these types of nutritional deficits cause an individual to crave foods their brain suspects will satisfy their dietary needs. Connections between nutritional deficiency and pica have also been offered up to explain the higher prevalence of pica in pregnancy, as pregnant women face increased nutritional demands and are more likely to lack important nutrients in their diet.

Nevertheless, the evidence doesn't point to a clear connection between pica and nutritional deficiencies—at least not in all cases of pica.[9] Instead, it seems likely that different cases of the condition have different (and sometimes multiple) causes. In some cultures, for example, pica appears to be a learned behavior that's passed down almost like a tradition.

In parts of the southern United States, it's a relatively well-known practice to eat clay (most commonly a white, chalky clay called *kaolin* and colloquially known as "white dirt"). One study of the black residents of a Mississippi county in the late 1970s—although, it should be noted, this practice is seen in white populations as well—found that 57 percent of women and 16 percent of children regularly ate clay, with an average daily consumption of 50 grams (about the weight of a typical Snickers bar). Mothers often passed the habit of clay eating down to their kids by soothing their young children with pieces of clay to nibble on.[10] And even though this study is over 40 years old, more recent studies have continued to note the practice in rural southern regions of the US.[11] A 2015 documentary, *Eat White Dirt*, focused on the subject, interviewing Southerners who openly discussed their predilection for eating clay. As one interviewee in the documentary stated, "For me a good day is a bag of white dirt and a Coca-Cola...I eat it every day."[12]

But some pica patients don't seem to be influenced by nutritional deficiencies or traditional practices. They complain their urges to eat unusual non-food items develop in their heads unbidden and become difficult-to-control obsessions. These obsessive thoughts lead to a compulsion—an irrepressible need to eat the item, even though it may seem obvious to the patient that consuming it is not in their best interest.

For example, one 10-year-old boy named Hamza was referred to a pediatric clinic after spending five years eating carpet fibers. Interestingly, Hamza did have iron deficiency anemia when doctors first examined him at the clinic, but his pica did not disappear even after his iron levels were brought back to within normal range with iron supplementation. Hamza complained that he didn't want to eat the carpet fibers, but he felt an overpowering urge to do so. As he tried to battle the urges, he experienced an intolerable build-up of tension; he would eventually succumb to the cravings and eat some carpet fibers to relieve the pressure.

But once Hamza was done eating the carpet fibers, it wouldn't be long before the urges began to bubble up inside him again—only to eventually reach another crescendo that would need to be alleviated by consuming more carpet fibers.[13] The description of

Hamza's struggles would sound familiar to many psychiatrists, as his symptoms resemble those of a commonly diagnosed psychiatric disorder: obsessive-compulsive disorder, or OCD.

Irrepressible urges

OCD is a condition that affects between 2 to 3 percent of the population at some point in their lives.[14] People with OCD are plagued by persistent, intrusive thoughts called *obsessions*. The obsessions typically lead to *compulsions*, which are acts a patient feels compelled to perform—usually to mitigate the anxiety and discomfort caused by obsessive thoughts.

For some, compulsions involve observable behaviors, such as handwashing to allay concerns about being exposed to some contaminant, or checking the dials on the stove multiple times to alleviate fears of a potential fire. In other cases, however, compulsions might consist primarily of mental acts like praying, reviewing past events, or counting. In many cases, the compulsions don't have a logical connection to reality. A patient might, for example, feel the compulsion to turn the lights on and off multiple times while thinking that failure to do so would cause harm to come to a family member. Most OCD patients appreciate how irrational this type of reasoning is, although that recognition typically doesn't make them feel any less compelled to engage in the behavior.

The term OCD has become familiar in the vernacular to describe someone who is very meticulous, but individuals who suffer from diagnosable OCD experience symptoms that are far more distressing and life-disrupting than just a tendency to want their desk organized in a specific way. Take the case of a 14-year-old girl named Amy, who suffered from severe OCD that centered around concerns about contamination—something many OCD patients become hyper-focused on. Indeed, close to 50 percent of people with OCD develop an intense concern about being exposed to dirt,

germs, toxic chemicals, etc.,[15] but typically their anxiety is disproportionately greater than any risk truly posed to them by these
things. Amy's OCD was concentrated on a specific type of contamination: she was afraid of being infected with pinworms.

If you know anything about pinworms, you'd probably agree
that Amy was justified in finding them particularly disgusting.
Pinworms are tiny (less than a half-inch long) white intestinal
worms. People sometimes unintentionally ingest or inhale their
microscopic eggs, which then travel through the gastrointestinal
tract and hatch in the intestines. After hatching, the new worms
mature over the next several weeks, then begin mating.

The next part is where it gets really gross. Once a female pinworm is pregnant and ready to lay her eggs, she patiently waits
until her host is asleep (how pinworms know their host is asleep is
still something of a mystery and perhaps one of the most disturbing things about them—and that's saying a lot). Once the host is
in a state of slumber, the female pinworm crawls out through the
anus, wriggles around the anal opening a bit, and lays thousands of
eggs (on average over ten thousand) in the folds of skin surrounding the anus. The eggs are covered in a sticky substance that helps
them adhere to the skin long enough to mature—so they are ready
to infect another host once they eventually become dislodged. A
side effect of that sticky material is that it irritates the skin around
the anus, often causing severe itching. This anal itching is the most
common symptom of pinworms; it's probably not helped by the
sensation of little worms crawling around your butthole.

The itchiness helps spread the infection, as people (mostly kids)
will scratch their anus, getting pinworm eggs on their hands and
lodged under their fingernails. From there, they might unintentionally deposit the eggs on furniture, bathroom fixtures, toys,
etc. When someone else touches one of these objects, they can get
eggs on their own hands and accidentally ingest them when they
put their hands to their mouth. Or—because pinworm eggs are
microscopic and incredibly light—they can get thrown into the air

(e.g., when bedsheets are rearranged to make the bed) and inhaled unknowingly. They also can be re-ingested by the infected person, which starts the process over in them anew.*

On top of it all, pinworms are not rare—they are the most common intestinal worm infection in the United States; it's estimated that over 10 percent of people in the US are infested with them.[16]

It's enough to make anyone's skin (or anus) crawl, but Amy took it to the extreme. She became obsessed with pinworm contamination and thought about it all the time. She developed compulsions that initially involved washing her hands repeatedly throughout the day to avoid the risk of infection—to the point where her skin was dry, cracked, and painful.

Amy's handwashing was just the beginning, though. Eventually, her apprehension of oral contamination caused her to be afraid to open her mouth; this led to a 10-month spell where Amy essentially did not speak. She then became concerned about accidentally ingesting pinworms while eating, so she refused to eat for four weeks. She was hospitalized for dehydration, and after 10 days in the hospital, a combination of medication and therapy helped her to begin talking and eating voluntarily again.[17] OCD is a chronic and persistent disorder, however, so it's likely that Amy's battle with the condition continued long after her hospital discharge.

The neuroscience of OCD

For decades, neuroscientists have been trying to understand what causes the unusual presentation of behavior we see in OCD. While the neuroscience of OCD is still not fully clear, most researchers believe a brain network that connects the *prefrontal cortex* with a group of structures called the *basal ganglia* is at least partially to blame.

---

* If there's any bright side, it's that pinworm infection can usually be easily treated with over-the-counter medications.

Unsurprisingly, the prefrontal cortex refers to the front of the brain—in fact, it's the part of the brain found at the very front of the frontal lobe. The prefrontal cortex makes up a large proportion of the entire brain—about 12.5 percent

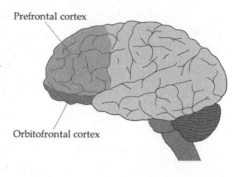

Prefrontal cortex

Orbitofrontal cortex

of its total volume.[18] Consequently, it's involved in a long list of functions. But it's best known for its role in higher cognition—the type of stuff that makes us distinct from many other "lower" animals, and includes functions such as sophisticated decision-making, judgment, impulse control, and rational thinking.

Prefrontal cortex circuits likely have multiple roles in the manifestation of OCD, but one part of the prefrontal cortex—the part that sits just above the eye sockets and is known as the *orbitofrontal cortex*—is thought to make an especially important contribution to the disorder. The orbitofrontal cortex is highly active when we notice something in the environment that we consider dangerous or threatening. For example, if someone with OCD who has fears of contamination accidentally touches a door handle in a public restroom, the neurons in their orbitofrontal cortex would start firing like crazy.

Some neurons in the orbitofrontal cortex leave the region and stretch to a group of structures called the basal ganglia, which are situated deep within the brain, near the bottom of the organ—hence the use of the word *basal*. (The term *ganglia* is plural for *ganglion*, a word that refers to a cluster of neurons.*) The basal ganglia

---

* It should be noted that the use of the word *ganglia* in this context is a bit of a misnomer. That's because the term *ganglia* is typically only used when referring to neurons outside the brain and spinal cord. Because the ganglia that make up the basal ganglia are found in the brain, they technically aren't ganglia. They would more accurately be considered *nuclei*. Regardless, *basal*

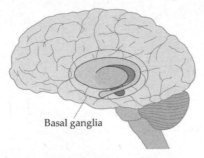

Basal ganglia

include a collection of brain regions, but to make things simpler I'll focus on the regions as a group rather than naming them individually. Each of the components of the basal ganglia has extensive roles of its own in the brain, but the regions also form a network that is critical to movement, cognition, emotion, and more. What is most relevant to our discussion here is that the basal ganglia are thought to be involved in the facilitation of goal-directed actions, the development of habitual responses, and the ability to switch to a different behavior when a current action is deemed unlikely to accomplish the goal at hand.

While the neural circuitry of the basal ganglia is complex, it is often simplified into consisting of two opposing pathways: the *direct pathway*, which facilitates action, and the *indirect pathway*, which inhibits it. Conceptualizing the basal ganglia circuitry in this way will help us to outline an easy-to-understand mechanism of what might be happening in the brain of an OCD patient.

When orbitofrontal cortex neurons are activated in response to a threat (real or imagined), they stimulate the basal ganglia's direct pathway to prompt an action to mitigate the threat. For our OCD patient who touched the door handle in the public restroom, this action might involve washing their hands or using an immoderate amount of hand sanitizer.

In a healthy individual, after accomplishing this action, the indirect pathway of the basal ganglia would kick in and inhibit further activity. But in someone with OCD, a few things can go wrong with this orbitofrontal cortex–basal ganglia connection. First, the orbitofrontal cortex and the pathway that connects it to the basal

---

ganglia is the accepted and most commonly used term for this group of structures, so that's how I'll refer to them.

ganglia tend to be highly excitable. This hyperactivity might cause OCD patients to be overly vigilant about things in the environment they consider a threat. In an OCD patient, for example, their orbitofrontal cortex might be activated not only when they touch the door handle in a public restroom, but also when they brush up against any surface that hasn't recently been sanitized—even if it's a countertop in their own home.

This extreme vigilance is associated with over-excitation of the direct pathway in the basal ganglia. Again, after a concern about contamination, activation of the basal ganglia direct pathway might prompt someone to clean their hands. But in a person with OCD, the extraordinarily high activity in the direct pathway drowns out the inhibitory action of the indirect pathway, and the patient has a difficult time switching to a different behavior. They get stuck, in a sense, in a habitual behavioral loop—like a skipping record.

Each time the behavior is performed and the threat is temporarily alleviated, the individual feels a transient sense of relief. This relief, however, compounds the problem because it reinforces the response (e.g., handwashing), causing the brain to link that action to a positive outcome. Thus, the brain is inclined to use that response again—and again and again—almost in an addictive manner.

This model of OCD has become popular in neuroscience, but at the same time modern researchers recognize it is a bit of an oversimplification. One problem is that the orbitofrontal cortex is not a homogenous brain region. While some areas of it are hyperactive in OCD, others appear to be underactive. Additionally, research has found that other parts of the brain (such as the *amygdala*—a region we'll discuss more in later chapters) also play important roles in the manifestation of OCD symptoms, suggesting the model outlined above is not quite complete. Finally, it seems likely that brain activity in OCD patients depends to some degree on the patient and their age and symptom profile. Nevertheless, many neuroscientists believe these pathways between the orbitofrontal cortex and basal ganglia are important in explaining the behavioral aberrations of OCD.

Obsessions and compulsions aren't only seen in OCD—they're core features of a variety of other disorders as well, which has led some scientists to suggest there is an "obsessive-compulsive spectrum" that includes OCD and a list of other conditions. Some argue, for example, that nail-biting (known medically as *onychophagia*) belongs on the obsessive-compulsive spectrum, and similar arguments have been made for *kleptomania* (compulsive stealing), *trichotillomania* (compulsive hair-pulling), compulsive sexual behavior, and more. But one particular disorder on the OCD spectrum has spawned an especially large amount of public interest, leading to the development of a popular television show.

Hoarders

When the television series *Hoarders* debuted in 2009, hoarding was already a documented phenomenon, but until 2013 it was typically considered a symptom of OCD and not a disorder in its own right. In 2013, however, the fifth edition of the *Diagnostic and Statistical Manual of Mental Disorders*, or DSM-V (a diagnostic guide commonly used by medical and mental health professionals in the United States), classified hoarding as a separate condition in the category of "Obsessive-Compulsive and Related Disorders."

A hoarding diagnosis applies to someone who habitually has extreme difficulty getting rid of possessions—even if those possessions have little to no actual value. Many of these patients (up to 95 percent of them)[19] also compulsively collect items—sometimes by purchasing them, sometimes by seeking out freebies. The hoarded items can be almost anything, from broken furniture to old newspapers to trash to pets. As the accumulation of items continues, the patient's living area fills with clutter, sometimes to the point where conditions become unsanitary and unlivable. As any viewer of *Hoarders* recognizes, these cases can be extreme—and even potentially deadly.

When Jesse and Thelma Gaston got married, they weren't hoarders—at least not to a degree that was apparent to anyone around them. Friends and family recall being invited into the Gastons' home early on in their marriage and nothing seemed unusual at the time. But over the years, the Gastons became more hesitant to allow others into their house. They started meeting visitors on the porch, and eventually only in the yard.

They had good reason to be reluctant about accepting visitors. Inside the house, the Gastons were building up a collection of old mail, clothes, garbage, etc., that piled from floor to ceiling in every room.

Eventually, trash began to accumulate in the Gastons' yard. Neighbors had been very tolerant of the Gastons—who at this point were in their 70s—but when the debris started to spread outside the house, neighbors complained. Then, the Gastons disappeared.

Several weeks went by. New mail piled up on the porch and parking tickets collected on Jesse's car. Something didn't seem right, and concerned neighbors called the police.

When officers came by to check on the couple, no one answered the door—but a vile smell was emanating from the house. Police decided it was imperative to take a look inside. Of course, the major concern at the time was that the couple had died inside their home, and their decaying bodies were producing the stench.

The police forced their way in and found garbage and debris piled so high and spread so extensively throughout the rooms that the only way to move around was by climbing over—and wading through—the trash. They realized they would have to dig through garbage just to uncover the Gastons' bodies. Eventually, however, they found the Gastons alive—buried in their own rubbish. Thelma had been trapped first when debris fell on top of her. Jesse had tried to rescue her but had become stuck in the process.

They had been trapped in the garbage for as long as three weeks.

Apparently, rats had attempted to take advantage of their immobility and began treating them as a snack, because they both had rat bites all over their bodies by the time they were rescued. Jesse died six weeks later of cancer—a demise likely precipitated by his harrowing ordeal. Thelma was too weakened to attend Jesse's funeral.[20]

An excess of pets

As I mentioned, some hoarders focus on animals rather than objects. These situations also frequently turn nightmarish; the animals become impossible to care for and are often neglected, undernourished, and diseased. The residence becomes contaminated with animal urine and feces, and when authorities intervene there are usually dead animals found decaying on the premises. Often, the hoarded animals are common pets, such as cats and/or dogs.

One case involved a woman who kept 92 cats in a 7.5-foot by 11-foot trailer—which amounts to less than one square foot of space for each cat. When animal control removed the animals, most of them were "covered in urine and feces," malnourished, emaciated, infested with fleas, and sick. Some were missing limbs; others were missing eyes. The owner of the trailer claimed she was keeping the cats to save them from being euthanized.[21]

In another case, complaints of a foul odor emanating from a residence brought animal control out to a home in Anchorage, Alaska. One of the enforcement officers stated that when he walked into the house, the smell of cat urine "literally burnt [my] throat." Debris and feces covered the floor, and there were somewhere between 180 and 200 cats in the home. Cats were everywhere, even in the ceiling.[22]

But animal hoarding is certainly not limited to cats and dogs. Glen Brittner began raising rats as pets after his wife passed away. He started with three rats, but they managed to get out of their cage and escape into the walls of Glen's home. As rats will do, they mated with one another and produced more and more—and more—rats. While most homeowners at that point would have considered calling an exterminator, Glen embraced the growing community of rats that was sharing his house with him. He fed them regularly (often just by tossing food onto the floor, where an eager mass of rats would quickly swarm on it), provided them with water, and let them procreate and spread throughout the house.

Eventually, Glen's house was full of rats. They lived in the walls, cupboards, and mattresses. They chewed on everything. Glen had to sleep outside in his shed; otherwise, the rats would pull his hair out in his sleep to use it in their nests. Sometimes they would even lick his eyes and lips for moisture while he was sleeping. The Humane Society removed over 2,000 rats from Glen's home, but afterward Glen found another 350 rats still living in his walls.[23]

The brain of a hoarder

Although we have yet to fully determine what makes the brain of a hoarder different from the brain of someone who is suffering from OCD without hoarding, there are some known characteristics that contribute to hoarding symptoms. Hoarders, for example, are often severely indecisive, which exacerbates their difficulties making a choice to get rid of things. More than a simple wishy-washiness, this difficulty in making decisions seems to represent problems with normal information processing. In other words, hoarders experience deficits in the cognitive capabilities necessary for efficient decision-making.

Hoarding patients also frequently display other cognitive issues such as difficulties with planning and problem-solving, and up to a fifth of them have diagnosable attention deficit hyperactivity disorder (ADHD).[24] On top of it all, they show poor self-awareness about their hoarding behavior, meaning that—as difficult as this may be for a non-hoarder to understand—hoarders often genuinely don't realize there's anything wrong with what they're doing.

One study of hoarders found they were unaware they were hoarding until 10 or more years after the symptoms started to appear.[25]

Studies that have looked at the brain function of hoarders point to aberrant activity in a number of brain areas, but—like the OCD patients we discussed earlier—many display abnormal activity in the prefrontal cortex. This atypical activity often occurs during tasks involving decision-making or impulse-control, and researchers propose it may contribute to the maladaptive attachments hoarders form with objects (or animals).[26]

Patients who have experienced damage to the prefrontal cortex and afterward developed hoarding behaviors support these conclusions. One man in his late 40s, for example, began collecting household electrical appliances—including televisions, vacuum cleaners, refrigerators, and washing machines—after surgery to remove a brain tumor. He accumulated 35 televisions in his living room, and when he ran out of space there, he started storing them in other rooms. Eventually, he was shoving TVs into ventilation ducts due to a lack of space. His wife coaxed him back to the doctor in search of an explanation for his new and unusual behavior, and brain imaging indicated damage to his prefrontal cortex—likely a consequence of the tumor and surgery.[27]

Although evidence supports the role of the prefrontal cortex in hoarding behavior, neurobiological research into hoarding is still in its early days. Neuroscientists hope to learn much more in the coming decades about what makes the brain of a hoarder different from that of a patient with typical OCD, as that information may help to guide the treatment of a disorder that can turn daily living into a cluttered, filthy—and potentially hazardous—situation.

Obsessions and compulsions are a reminder of how easily we can lose control of our brains. In disorders characterized by obsessions and compulsions, patients complain they are tormented by their thoughts and compelled to engage in behavior that ranges in severity from annoyingly undesired to traumatically distressing.

These thoughts and behaviors are unbidden and unwanted; it's almost as if a malicious provocateur has situated himself in the brain, manipulating our behavior like a malevolent puppet master. One day perhaps neuroscience research will shed enough light on obsessive-compulsive disorders to enable those who experience them to wrestle back control of their thoughts and lives.

For now, let's have a bit of a palate cleanser. So far, we've primarily been discussing disorders that point to deficits in how the human brain works. But are there ever things that go "wrong" with the brain that cause it to work *better*? It sounds improbable—almost like injuring your foot and then being able to run faster. Nevertheless, there are cases where atypical brain development—and even brain damage—unlocks hidden talents, leading to a type of higher-level function that can seem almost impossible to achieve with even the most dedicated practice regimen. These cases have raised questions about the true potential of the human brain.

4

EXCEPTIONALISM

No one gave Kim Peek much of a chance early on in life. He was born in 1951 with a head so large his neck couldn't support it, and he had a condition known as an *encephalocele*, where an incompletely developed cranium allows part of the brain to bulge outside the skull—potentially twisting, distorting, and damaging brain tissue in the process. Kim's encephalocele consisted of a sac-like protrusion about the size of a baseball. It was a life-threatening situation, but survival also brought with it a high likelihood of lasting intellectual and/or physical disability.

When Kim was 9 months old, his chances of surviving were becoming more encouraging, but some long-term consequences of his early-life complications were already apparent. One particularly insensitive doctor told Kim's parents that Kim was "retarded" and suggested he should be institutionalized so his parents could "get on with [their] lives."[1] Kim was, in the eyes of the mostly unsympathetic medical field of the day, a lost cause. Kim, however, would go on to develop some of the most extraordinary intellectual abilities ever documented in a human being.

Kim's encephalocele spontaneously resolved at around age 3 but left considerable brain damage in its wake. This led to pronounced delays in Kim's physical development. He couldn't walk on his own until he was 4 years old. Before then, he crawled around with his large, unsupportable head dragging on the ground. Even after he started walking, he had an unsteady, awkward gait, and he remained unable to walk up and down stairs on his own until he was 14. Throughout his life, he had difficulty with fine motor skills and needed assistance in completing daily tasks such as brushing his teeth, buttoning his shirts, and combing his hair.

Kim's intellectual and social development were also abnormal. As an adult, he was recorded as having a below-average IQ of 87, with some of the scores on his IQ subtests falling into the range of intellectual disability.[2] He was severely introverted and could not look someone in the eye until well into adulthood. At the same time, he was hyperactive and had difficulty inhibiting inappropriate social behavior. He lasted less than 10 minutes in first grade before his teacher asked him to leave due to his incessant commentary, which was often directed at no one in particular.

Despite all of Kim's developmental abnormalities, his parents began noticing some extraordinary things about him in the early years of his life. When Kim was 3, he asked his father, Fran, what the word *confidential* meant. Fran facetiously told Kim to go look it up. Kim did just that; he crawled over to Fran's desk, pulled himself up, and found the definition in the dictionary entirely on his own.[3] Kim's parents had never taught him to read or comprehend alphabetical order, but in accomplishing that task he demonstrated an understanding of both.

By the time Kim was 6, he was displaying prodigious memory capabilities. He was able to recite complete paragraphs from books he had read after simply being given a page number, and he had memorized the entire index for a set of encyclopedias.

As Kim got older, he read everything he could get his hands on—from biographies to atlases—memorizing much of what he read in the process. He also became obsessed with numbers; sometimes, he perused phone books just to add the telephone numbers down the page. Although he still lagged in other physical and cognitive abilities, his unique mnemonic and arithmetical faculties flourished. Additionally, he acquired amazing speed-reading skills. He eventually could read a page in 8 to 10 seconds while memorizing all the information on it. He even began reading and comprehending the right and left pages of a book simultaneously (with his right and left eyes).

By the time he died in 2009 at the age of 58, Kim had read—and *memorized*—more than 12,000 books. This led to encyclopedic knowledge in a wide range of subjects, including American history, geography, literature, classical music, sports, movies, and more. He memorized all the area codes in the country and all the major zip codes in the United States and Canada. And he had an amazing knack for calendar calculations; if you gave him a date, he could (often in less than a second) tell you the day of the week it fell on—along with other details, such as the occurrence of major holidays.

And yet, Kim's brain had some abnormalities that any neurologist would consider likely to handicap—not enhance—cognitive performance. For example, it was completely missing the largest collection of nerve fibers in the brain—a structure known as the *corpus callosum*. The corpus callosum acts as a vital avenue of communication between the two cerebral hemispheres. (We will discuss the corpus callosum further in Chapter 11 when we look at other patients who have experienced damage to the structure.)

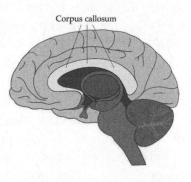

Corpus callosum

Although it's rare, sometimes children are born missing all or part of their corpus callosum. This condition, known as *agenesis of the corpus callosum,* can result in highly variable symptoms. Some individuals experience developmental delays, seizures, intellectual deficits, and/or visual or auditory impairments. But others never have any discernable problems; there may not be any indication they are missing the largest neural pathway in the human brain.

This creates a bit of a conundrum for neuroscientists. It's a challenge to explain why damage to a prominent brain structure could be debilitating for one patient but hardly noticeable for another. And it's more difficult still to understand how, in someone like Kim Peek, the lack of a corpus callosum might be linked to such extraordinary abilities.

Some have proposed that Kim's brain was forced to develop alternate pathways of communication between the two cerebral hemispheres to make up for the lack of the corpus callosum. And maybe those alternate pathways were responsible for causing Kim's brain to work in such an exceptional way. On the other hand, perhaps Kim's missing corpus callosum played little part in his cognitive abilities, and it was a coincidence that this structural abnormality was present in someone with his remarkable talents.

In truth, we may never know. Because Kim was so extraordinary (and because he is no longer living), we are limited in our ability to draw conclusions about his skills. Typically, if neuroscientists wanted to study a special presentation of abilities such as Kim's, we would get a group of similar individuals together and use neuroimaging—an approach that provides details about the structure and/or function of a living brain—to examine their brains. Then we would compare their brains to those of people without such talents, looking for common themes. If all people with exceptional skills had agenesis of the corpus callosum, for example, it would suggest that this anomaly is important for the development of unique capabilities like Kim's.

But Kim Peek was one-of-a-kind. No one else has displayed his skillset to the degree that he did. And because his faculties aren't apparent in anyone else who lacks a corpus callosum, it suggests there must have been some other neurological aberration involved in the development of his abilities. Although Kim did have several other brain abnormalities, none of them seem capable of fully explaining his talents.

Extraordinary people

Kim Peek was a *savant*, a French word that means "a knowledgeable person." In a medical context, the word *savant* describes someone who displays great skill or talent in one specific area, despite being impaired in other ways—often due to a developmental disability, brain injury, or brain disease. The term was first used in this context by Dr. J. Langdon Down (best known for writing the original description of Down syndrome) in the late 1800s.*

Interestingly, the talents savants display have common themes. They usually can be grouped into one of several categories: music (e.g., having perfect pitch, piano playing), art, calendar calculations, mathematics (e.g., making rapid calculations), and mechanical or spatial skills (e.g., estimating distances with extreme accuracy, having an intuitive sense of direction). Typically, whatever a savant's special faculties, he or she also displays exceptional memory capabilities.

Savants have skills that are especially impressive in contrast to their disability. But some, referred to as *prodigious savants*, have genius-level talents that would be incredible even in someone with no other impairments. Kim Peek, of course, was a prodigious savant.

The condition savants experience is called *savant syndrome*, and it underscores how much we have yet to understand about the brain. For despite all our advancements in neuroscience, we have minimal comprehension of how abilities like Kim's—which to some degree seem miraculous—can manifest in anyone, much less someone with a substantial developmental disability.

Of course, however, hypotheses about what causes savant syndrome do exist. They are preliminary but based on a growing body

---

* The term Down initially coined was *idiot savant*. At the time, the word *idiot* was commonly used to refer to individuals with an IQ below 25, but others since Down have noted that savants usually have an IQ above 40. Therefore, the term idiot savant is inaccurate, and of course the word *idiot* is considered demeaning today, making savant the preferred term.

of research into the savant brain. Surprisingly, most hypotheses suggest that savant skills emerge by inhibiting—not enhancing— functions of the typical brain.

A brain without limits

The brain is adept at managing large amounts of incoming sensory information efficiently. It filters out the data it deems unimportant—keeping much of it from reaching our conscious awareness—and instead of forcing us to evaluate every sensory stimulus as if it were completely novel, it uses past experience to make assumptions about what we are perceiving. For example, suppose you are walking along the sidewalk and see a large, oblong object in the distance moving down the street at a relatively high speed. Your brain doesn't need to wait until the object is close enough to make out all the details before it draws on its conceptual knowledge of oblong objects that travel on roads and decides this one is likely to be a car. Instead, it applies that label early on, and it's right most of the time.

The brain's use of concepts and labels to make sense of the world has distinct advantages. It can, for example, speed up how quickly you're able to assess what's going on in your environment, and it can accelerate learning by grouping things into categories. At the same time, using previously developed concepts and making predictions based on those concepts is not without its shortcomings. It causes our view of things to be highly influenced by our past experiences, making us more likely to misunderstand or overlook novel stimuli.

But what would happen if the brain did not attempt to categorize and apply labels to all incoming information? Instead, what if more raw sensory data were accessible to our brain? At first, this onslaught of sensory information would likely be overwhelming. It might make someone highly sensitive to novelty and cause disruptions to normal cognitive function.

But the availability of more details about the world around us also might promote a perspective that isn't normally attainable to others. This accessibility of information could, for example, allow for greater perception of detail in a scene that someone could later re-create through painting. Or it might enable one to focus on individual elements without the brain attempting to categorize them—a tendency that could lead to an enhanced ability for rote memorization of the facts. Perhaps it would promote creativity, which often seems to be hampered by the proclivity of the human brain to see the world through the limitations of preconceived labels.

Thus, some researchers believe the savant brain is a brain that doesn't feel the need to categorize and apply labels to all information that comes into it. This provides more access to raw incoming data and could let someone reap the benefits that come along with that increased access. According to this perspective, the key difference between the savant brain and a typical brain is reduced activity in brain regions involved in concept and category formation and application.

Research suggests that one part of the brain that is important for categorization and concept formation is found in the left cerebral hemisphere, specifically in the anterior (frontal) part of the temporal lobe. Therefore, scientists have hypothesized that inhibition of activity in this region might lead to the types of skills seen in savant syndrome.

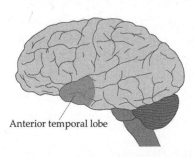
Anterior temporal lobe

Intrigued by a potential connection between the anterior temporal lobe and savant syndrome, researchers have attempted to induce savant-like skills in people by artificially inhibiting activity in their anterior temporal lobe using a method called transcranial magnetic stimulation, or TMS. A TMS device generates magnetic fields that penetrate the scalp and create transient electrical currents in the brain. These electrical currents aren't the type that cause a shock, but they can briefly influence the activity of neurons and thus can alter brain function. The effects of TMS are typically short-lived, but the technique has found some potential therapeutic uses, such as in treating depression, migraines, and obsessive-compulsive disorder.

TMS is commonly used in research to temporarily disrupt activity in a certain part of the brain. Researchers can then observe the effects of this disruption on an individual's experience and/or behavior. If behavior is influenced in a specific way after altering function in a particular part of the brain, it's logical to assume that the brain region that was affected plays an important role in the behavior (or lack of behavior) being observed.

When researchers applied TMS to the left anterior temporal lobe of participants who previously didn't display any savant abilities, they saw interesting results. Some of the subjects in these studies showed improvements in drawing skills,[4] detecting small errors in writing samples,[5] rapidly estimating the number of objects present in an image,[6] and forming accurate memories.[7]

I should point out that these studies did not result in astonishing changes. In other words, many of the research participants displayed improvements in the skills mentioned above, but they didn't develop savant-like abilities. There were enough changes, however, to prompt the experimenters to suggest that we all might have the potential for savant skills hidden inside our brains, waiting to be unlocked. But there is perhaps even more convincing evidence that people possess latent savant abilities: the sudden appearance of savant syndrome after a brain injury.

Sudden talent

In 2006, Derek Amato was nearing 40 and had yet to find stability in his life. In the previous 20-odd years, he had started his own pressure-washing business, sold cars, did public relations for a nonprofit organization, taught karate, and delivered the mail, among other things. After leaving yet another job, Derek decided (as so many of us do when we are feeling a bit aimless) to visit his hometown.

So, Derek traveled back to Sioux Falls, South Dakota, to visit his mother. While Derek was in Sioux Falls, he met up with some old friends for swimming and a barbecue. Derek's life would be dramatically changed by the end of that get-together.

Derek and his friends were playing catch with a football around the pool, and at one point Derek called for the ball so he could make a catch while diving into the water. Derek leaped into the pool, and his friend threw the football while Derek was in midair. Derek made the catch but miscalculated the depth of the shallow end. He plunged into the water and hit his head on the bottom of the pool.

Derek emerged from the water with his hands clutching his head. He needed help just to get out of the pool. His friends were talking to him, but he couldn't hear anything. He knew right away that he had suffered a serious injury.

Doctors told Derek he had a severe concussion and that he might have long-term hearing loss. There would, however, be other lasting effects. From that point on, Derek was plagued by intense headaches, memory problems, and extreme sensitivity to light. But he also developed a new talent that even the most astute physician could never have predicted as a consequence of his brain trauma.

Several days after the accident, Derek stopped at his friend Rick's house to say goodbye before leaving to drive back to his home in Colorado. While talking to Rick, Derek noticed a small electric keyboard on the other side of the room, and he felt inexplicably drawn to the instrument. He had an irrepressible desire to play it, even though he had never taken piano lessons (and had never been particularly interested in playing the piano before).

When Derek turned the keyboard on and started to play, something magical happened. His fingers danced across the keys with the fluidity of a professional pianist. He composed a new piece of music on the spot, smoothly interweaving chords and notes that had—up until that moment—been unknown to him. He played for the next six hours. His friend Rick was amazed. "I couldn't believe what I was hearing," Rick said. "I was just blown away."[8]

On that day, Derek found his true calling—even if it was foisted upon him by a traumatic brain injury. He would devote his life to music.

Since his accident, Derek has toured the country as both a solo act and an accompanying pianist. He has recorded two albums and appeared on *The Today Show, NPR*, and various other media outlets to perform. Derek has discarded his previously haphazard existence and found a purpose through his new talent: he is a musician with a passion for displaying the gift that he feels was bestowed on him by God.

Out of the blue

Derek has *acquired savant syndrome*, a condition that involves the sudden development of exceptional skills—typically after brain injury or disease. Acquired savant syndrome is incredibly rare, with just over 30 documented cases (although it's uncertain how many undocumented cases may exist).[9] But each case is as astonishing as Derek's.

Jason Padgett, for example, was working in a futon and furniture store, living a life that "rotated around girls, partying, drinking, waking up with a hangover, then going out and chasing girls and going to bars again."[10] He was not someone you would expect to display any genius abilities. It took him two tries to complete his senior year in high school. He attempted community college but dropped out.

In September 2002, Jason was attacked and mugged outside a karaoke bar and suffered a severe concussion. Then, his priorities and perspective on life suddenly shifted.

It all started with a change in perception. The world around him appeared pixelated; he saw lines and geometric shapes in every-thing ranging from flowing water to sunlight. He began to spend hours and hours drawing fractals (complex geometrical images that consist of repeating patterns). For Jason, however, these were more than just drawings—they were visualizations of complicated relationships in mathematics and physics. It wasn't long before he had over a thousand drawings of fractals and other elaborate mathematically themed images.

Jason's new structural, mechanistic way of seeing the world led him to develop an obsessive interest in mathematics and physics, simply because the tenets of these disciplines aligned best with his new understanding of how the world worked. Soon, Jason was pondering complex questions, such as how to determine the nature of spacetime. He discovered he had an intuitive understanding of sophisticated mathematical principles, and he is now considered by some to be a mathematical genius.

Orlando Serrell was living the life of a typical 10-year-old boy when he was hit in the head with a baseball during a baseball game in 1979. The blow knocked Orlando to the ground, but after a short time he was up again and back in the game. He had a lingering headache but didn't seek medical attention. Soon after the inci-dent, however, Orlando noticed he could rapidly determine the day of the week for any date after his blow to the head. In other words, you could say: "September 25, 1985," and Orlando would immediately respond by telling you, "Wednesday."

Orlando claims he doesn't even have to think about the calen-dar calculations; they pop effortlessly into his head. He has also developed an unusually precise memory for the events of his life since the accident. For every day since his accident occurred, he can recall what the weather was, and for many days he can remem-ber exactly what he was doing. He only needs to be provided a date for these details to materialize in his mind.

Other acquired savants have developed their abilities after stroke, dementia, brain surgery, or some other brain insult. But just in the past decade, scientists have realized that damaging effects on the brain aren't a prerequisite for the development of new savant-like skills. Over a dozen cases of acquired savant syndrome have now been documented in patients with no prior insult or injury.[11] These individuals developed their savant skills out of the blue—much to their amazement and the surprise of those around them. Researchers have termed them *sudden savants.*

Michelle Felan, for example, woke up in the middle of the night in December 2016 with an irrepressible need to draw. She was 43 years old and had never shown any interest in art, nor did she have any artistic training. But she stayed up the rest of the night with, in her words, "a compulsive need to draw, which continued over the next three days at an intense level."[12] Within three months, she had created

A piece of artwork by Michelle Felan titled "The Mayan." Reproduced with permission. Copyright 12/31/2016, Michelle L. Felan.

15 pieces of art that—at least to my amateur eye—have the appearance that they were produced by a professional. She now spends eight hours a day working on her artwork, and some have compared her style to Frida Kahlo and Pablo Picasso.[13]

Thus, there seem to be no clear rules to savant syndrome. The available evidence (as scant as it is) suggests it can potentially occur in anyone, at any time, with no precipitating event. Perhaps the most consistent underlying theme is that the savant behavior tends to manifest itself as a compulsion. Savants often feel like they *must* create their art, perform their calculations, memorize facts, and so on, because they are being compelled by a force that cannot be denied.

Because savant syndrome is so rare, it has been challenging to study the neuroscience of the condition. To confidently draw conclusions about it, neuroscientists would need to examine a large number of savant brains to look for commonalities. But due simply to a lack of cases, most studies of the savant brain include only one savant. While informative, this approach may do more to identify the eccentricities of an individual savant brain than to tell us about the general neuroscience of savant syndrome.

Thus, there are very few conclusions we can draw about savant syndrome, whether someone is born with it or acquires it later in life. But, like many of the other conditions discussed in this book, savant syndrome raises fundamental questions about brains and the human experience. For, if savant capabilities can emerge suddenly in someone after a head injury—or for no apparent reason whatsoever—then does it suggest that these capabilities are hidden in all of us, just waiting for something to draw them out? Are human beings capable of much more than we currently even dream of achieving?

Perhaps answers to questions like this will emerge from future research with savants. Until they do, however, the fact that they remain unanswered severely diminishes our ability to claim that we have a deep understanding of how the brain works.

5

INTIMACY

Erika Eiffel used to be a world-champion archer. She was on the
United States National Archery Team in the early 2000s and
was once one of the top-ranked archers in the world with a compound bow. But a love interest got in the way of her archery career.

When Erika's new relationship began in 2004, her friends and
family felt Erika was squandering her talent to pursue a partner
who hadn't shown any signs of reciprocating her affection. Indeed,
it was hard to argue against the allegation that Erika's relationship
was extremely one-sided. Erika constantly fawned over her new
love interest, but no one ever saw the object of her affection display
any emotion toward her.

Her partner had no way of expressing emotions, however,
because it is an inanimate structure. Erika's romantic relationship
was with the Eiffel Tower.

Yes, you read that right. Erika was in love with the prominent
wrought-iron landmark that extends more than 1,000 feet into the
air above the Champ de Mars in Paris. And this was not just a pass-
ing fling; Erika LaBrie changed her name to Erika Eiffel around the
time she expressed her undying commitment to the Eiffel Tower in
a marriage ceremony in 2007. Although her marriage was not recognized by any law or convention, it seemed to Erika a fitting way
to commemorate her dedication to the structure.

This wasn't Erika's first relationship with an inanimate object.
She once had a long-term affair with an archery bow she called

Lance; she claims this union was a driving force in pushing her toward her archery achievements. She also had previously been discharged from the US Air Force Academy because of her attachment to a Japanese sword.

Erika's relationship with the Eiffel Tower, however, did not last—an outcome that Erika blames mostly on the media and the unsympathetic nature of public opinion. Erika's affair became highly publicized after the release of a 2008 British documentary called *Married to the Eiffel Tower*. Although she had agreed to be in the film, Erika felt the final version focused disproportionately on the sexual aspect of her bond with the tower. Afterward, Erika said she couldn't visit the Parisian landmark without being scrutinized by the staff and visitors. Heartbroken, she decided to move on from the relationship. Erika, however, found solace in a past romance: a 20-year love affair with the (at this point mostly destroyed) Berlin Wall.

Object of desire

Erika is a self-identified *objectum sexual*, a term used to describe those who display a predilection for romantic, emotional, and/or sexual interest in inanimate objects—an orientation also known as *objectophilia*. Although objectophilia is rare, dozens of people throughout the world claim to be objectophiles. The focus of their affection is widely varied. Some, like Erika, have fallen in love with famous landmarks, but examples of other things objectophiles have formed relationships with include: a car, an amusement park ride, a video game character, a statue of a Greek god, a body pillow with an image of an anime character on it, a fence, an electronic soundboard, and a sex doll.

To say objectum sexuality is poorly understood would be an understatement. In fact, it's a topic that's barely touched on in the medical literature, and very few studies have addressed it. Some see objectum sexuality as a perversion, but objectophiles argue that it's a sexual orientation just like heterosexuality or homosexuality—and something they have little control over.

Some professionals agree and classify objectum sexuality as a rare type of sexual orientation.[1] Considering it as such, it doesn't seem appropriate to talk about objectum sexuality as a disorder any more than any other sexual orientation would be considered a disorder. At the same time, clearly something is unique about the way objectum sexuals view love and attraction. While scientists are far from determining what makes the objectophile brain so extraordinary, they have found some clues that eventually might help to explain objectum sexuality.

One is that objectum sexuals report high rates of a phenomenon known as *synesthesia*. Synesthesia is a perceptual experience in which one sensation involuntarily elicits another. A common example of synesthesia is the association of sounds with specific colors. To a synesthete, a trumpet blaring might sound red, while a dulcet flute might have a clear connection to blue. Some synesthetes may even see these colors when they hear the associated tones.

There are, however, many different types of synesthesia. The linking of sounds with colors is known as *chromesthesia*, but another common form of synesthesia is *grapheme-color synesthesia*, where individuals involuntarily see letters, numbers, or words as being tinged with a specific color (M, for example, might appear blue, while A appears red). To some synesthetes, known as *grapheme-personification synesthetes,* these colors are also linked to perceived "personalities" of the letters, numbers, or words. A, for instance, might seem to be easily angered, hence the association with red (which is often correlated with anger even in non-synesthetes).

Synesthesia can also involve other sense modalities. Some people, for example, experience tastes that seem inextricably linked to certain words—a condition known as *lexical-gustatory synesthesia*. For James Wannerton, a British man with lexical-gustatory synesthesia, every conversation elicits an uncontrollable flood of tastes. When he worked at a pub, he often had to give change, which brought on an overwhelming taste of processed cheese. James claimed the name Derek (that of a regular customer) would inevitably elicit the taste of earwax,[2] although that assertion does raise the question of why earwax was such a familiar taste for James (I'll grudgingly admit to smelling earwax before, but I don't believe I've ever tasted it).

A study of objectum sexuals found that they experience synesthesia more frequently than others in the general population.[3] In addition to some of the more common forms of synesthesia mentioned above, however, objectophiles are more likely to experience *object-personification synesthesia*, where the brain automatically imbues inanimate objects with personality traits. This, of course, would make sense to occur in an objectum sexual brain. Objectophiles may have grown up associating personalities with objects all their lives, so it's not as much of a stretch for them to feel like they can socially interact with objects as well.

Nevertheless, most others who experience object-personification synesthesia do not also develop a love interest for objects. There must, then, be something else occurring in the brains of objectum sexuals that causes them to focus their romantic attention on things rather than people, and that additional mechanism remains a mystery.

In the eye of the beholder

Of course, sexual arousal from things that are not a typical part of a sexual relationship is relatively common. We normally call this type of interest a *fetish*.*

It's difficult to determine just how common fetishes are, as

---

* In medical parlance, the term *fetish* is typically only used when the interest becomes disruptive or distressing to the individual, but the word is used colloquially to refer to almost any unusual or especially intense sexual interest.

people are generally reluctant to divulge information about their most secretive sexual habits. One 2007 study attempted to shed some light on the popularity of specific fetishes—while also circumventing people's hesitancy to answer questions about their sexual interests—by exploring how often particular fetishes were the focus of online discussion groups (Yahoo! groups specifically).[4] They found an abundance of atypical sexual interests, and no shortage of people willing to talk about them (usually anonymously) in an online forum. Their search for Yahoo! groups that contained the word *fetish* either in the group title or description returned 2,938 groups with more than 150,000 members* (for the sake of comparison, a search for "basketball" found 3,471 groups).

Out of the many fetishes discussed in these groups, the most common was *podophilia*, or a sexual interest in feet. Other groups focused on body parts that are frequently targets of sexual attention, such as lips, legs, buttocks, breasts, and genitalia. But less popular groups also mainly discussed nails, noses, ears, and teeth.

Some groups were centered around articles of clothing that often have erotic associations, such as stockings, skirts, and underwear. But other groups focused on wristwatches, jackets, and even diapers. A surprising 933 people belonged to groups that talked about the attractiveness of stethoscopes, and 150 people were in groups that celebrated the allure of hearing aids. Twenty-eight people found catheters sexually appealing enough to join a group devoted to discussions of them.

Still other groups were fixated on stimuli that aren't just unusual, but typically considered more off-putting—at least in a sexual context. Eighty-two people belonged to groups that talked about an attraction to body odor. And, at the most distasteful end of the spectrum (although clearly that assessment depends on your tastes), 8,367 members belonged to groups that proclaimed their

---

* This number is likely inflated, as some individuals may have been subscribed to multiple fetish-oriented groups. The researchers estimate, however, that their study still likely included "many thousands" of participants.

sexual interest in bodily fluids and excretions, including urine, blood, mucus, and feces.

Penny jars, feet, and safety pins: exploring the neurobiological basis for fetishes

No one knows for sure how fetishes develop. Over the years, scientists have offered up many hypotheses; some are plausible, some less so. In the 1920s, Sigmund Freud proposed what may be the best-known—and probably the most bizarre—explanation for fetishism.[5] Freud suggested that a fetish (primarily seen by Freud as a male tendency) was an attempt to cope with a boy's traumatic realization that his mother didn't have a penis. This discovery, according to Freud, is extremely distressing because it provokes a fear that the boy's own penis is in danger of being removed (possibly by his father as a punishment for the boy's sexual attraction to his mother).

To cope with this trauma, Freud hypothesized, the child redirects the sexual interest he has in his mother's genitalia onto something else. The new object of his desire might be linked to one of the last impressions he had before learning his mother was lacking a penis, which would commonly include things such as feet (assuming he was seeing her genitals from a lower perspective), legs, underwear, etc. The object of his fetishism could act as a healthy outlet for his sexual interest in his mother and consequently help him avoid castration by his father.

Although Freud made some very important contributions to understanding how the mind works, under a present-day lens some of his ideas sound a bit half-baked; his view on fetishism is not something many psychologists still subscribe to. Nevertheless, childhood experiences remain a popular way of explaining the origin of fetishes.

According to one perspective, an object may play a role in a memorable sexual experience early in life, causing the brain to form a persistent mental association between that object and sexual gratification for years to come. That enduring memory could be the basis for a fetish.

Researchers have attempted to replicate this process in experimental studies, with some success. A 1999 experiment, for instance, investigated whether researchers could "train" a small group of men to become sexually aroused by a picture of a penny jar[6]—yes, a *penny jar*, an object presumably chosen for its unlikeliness of ever playing a leading role in anyone's sexual fantasies. To measure arousal, the scientists used a device referred to in their published report as a *Type A penis gauge*. Maybe I just have an immature sense of humor, but I can't help but chuckle thinking about a starchy academic experimenter uncomfortably asking an assistant to hook up the Type A penis gauge.

An apparatus like the Type A penis gauge is more generally known as a plethysmograph, and it typically consists of flexible metal bands that are placed around the shaft of the penis. The bands are connected to a recording device that makes note of changes in penile circumference, measuring what in scientific parlance is called penile tumescence, or the degree of blood flow into the penis. Blood flow into the penis is the cause of an erection, and thus penile tumescence acts as an indicator of sexual arousal. So, for future reference—in case it ever comes up in your personal or professional life—a Type A penis gauge is used to measure just how turned on someone is.

The participants in the experiment were nine males from undergraduate psychology classes at the University of North Dakota; they received extra credit and $20 for taking part in the experiment—which is probably not enough to compensate for the unease caused by learning you could get sexually aroused by a penny jar.

Researchers connected the Type A penis gauge, then showed participants pictures of attractive naked women immediately before or after images of the penny jar. Following just a few experimental sessions of pairing the sexually explicit images with the image of the penny jar, the image of the penny jar alone elicited significant increases in penile tumescence/sexual arousal. Thus, the participants had "learned" to find something that is about as nonsexual as it gets to be sexually arousing based on its association with some erotic pictures. This experiment says a lot about just how primitive our sexual brain can be, but it also may tell us something about fetishes, as they might form in the same way—through strong memories that link a previously neutral stimulus to a sexual experience.

Although the results are interesting, they leave a lot of unanswered questions. They don't tell us, for example, how persistent these types of learned associations are. This is a critically important question to answer if we think fetishes may emerge from associations made early on in life, but no studies have adequately explored it. And there are many fetishes in which individuals can't recall clear memories linking their fetish to a pleasurable experience. These certainly may be cases of impaired recall, but it's also possible other factors were involved in the development of their fetishes.

Some researchers, such as the well-known neuroscientist V.S. Ramachandran, have focused on these other factors. Specifically, Ramachandran has posited neurobiological mechanisms behind the development of fetishes. He suggests that foot fetishes, for example, might be caused by cross-wiring in the primary somatosensory cortex (the area of the brain discussed in Chapter 2 that processes touch sensations from the body).

Ramachandran's hypothesis is based on an anatomical feature of the primary somatosensory cortex: different areas of the primary somatosensory cortex receive information from different parts of the body. In fact, the primary somatosensory cortex can be partitioned into a map of sorts, divided into regions that receive information from distinct body areas. A specific part of your primary somatosensory cortex is activated when you touch your face, another is activated when you touch your leg, and so on.

Ramachandran developed his hypothesis about foot fetishes

based on the observation that the part of the primary somatosensory cortex that receives sensory information from the foot sits adjacent to the part that receives information from the genitals. Thus, it would take only a slight wiring error during early neural development to cause sensory stimulation of the feet to also stimulate the part of the primary somatosensory cortex that is linked to genital regions—perhaps

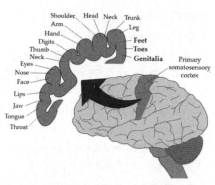

A rough map of the different parts of the primary somatosensory cortex that are devoted to processing information from different parts of the body. Note that the regions devoted to the feet, toes, and genitalia are all next to one another.

in a way that would generate sexual excitement. This slight miswiring, Ramachandran contends, could cause the feet to attain sexual relevance in one's brain.[7]

In support of a neurobiological perspective on fetishes, there have been cases where fetishistic behavior has either appeared or disappeared along with changes to the brain—suggesting a neurological origin. For example, one 38-year-old man named Henry had an unusual fetish for safety pins. When staring at a safety pin, he experienced enjoyment he described as "greater than sexual intercourse."[8]

Henry's attraction to safety pins started as a child. From that time on, he would often seek out a private place such as the bathroom to hide away and stare—yes, just stare—at a safety pin. When he was young, however, Henry also began to experience seizures while staring at safety pins.

It's important to note that not all seizures look like what you might picture in your head when you think of a seizure. Most people envision someone falling to the ground, losing consciousness, and experiencing repetitive muscle contractions that cause their body to convulse violently. That particular presentation of symptoms is called a *tonic-clonic* (formerly known as a *grand mal*) seizure, but other types of seizures can involve periods of staring off into space, muscle contractions in just one part of the body, or general confusion, among other symptoms.

For Henry, his seizures would begin when he started to get glassy-eyed while staring at a safety pin. Then, he would involuntarily make a low humming noise, followed by sucking movements with his lips. Finally, for a couple of minutes he would become immobile and unresponsive. The seizure would end as abruptly as it started, although he would often be confused for a period afterward. Sometimes he dressed in his wife's clothing following a seizure, a point noted in the published report of Henry's case without any additional explanation.[9]

It seemed that Henry's seizures were linked to the intense emotion brought on by safety pins (his seizures always started with either staring at a safety pin or imagining doing so). Still, he refused to give up admiring safety pins—even though he was having seizures every week.

Unable to control his seizures any other way, doctors eventually decided brain surgery might be the only option for Henry.

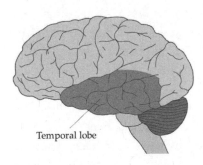

Temporal lobe

Using a device that can monitor electrical activity in the brain, they determined that Henry's seizures were originating from a region of the temporal lobe. Then, they surgically removed that part of the temporal lobe, in the hope that this would put an end to the seizures.

After Henry's surgery, his seizures did go away. But surprisingly, so did his safety pin fetish. When he came in for a follow-up visit about a year after his surgery, he told doctors he no longer had any desire to gaze adoringly upon safety pins. Instead, he had become much more interested in his wife again. I assume she was happy with this, although many women would harbor some resentment about being considered the second-best option behind a safety pin for much of their marriage.

Henry's case is certainly unique, but it does suggest that at least some fetishes can be attributed to atypical brain activity. For, if changing the brain alters a behavior, then the behavior must have been influenced by the brain in the first place. And, although the research on fetishes is somewhat scarce, if we expand our view to include all types of unusual sexual behavior, there are even more examples that enable us to trace such behavior back to neural circuitry.

The bewildering range of human sexual interests

Fetishes are just one type of *paraphilia*, a term that refers to any unusual sexual interest. But there are *many* paraphilias. Some are innocuous and may even be relatively common and accepted. Examples include *mazophilia,* which involves a preoccupation with female breasts, or *gynandromorphophilia*, an attraction to transgender women. Other paraphilias seem mostly harmless but are nonetheless exceedingly strange. *Dendrophilia* is a paraphilia where people find trees sexually arousing. *Formicophilia*, as described by sex researchers John Money and Ratnin Dewaraja, is a paraphilia in which someone is turned on "by small creatures like snails, frogs, ants, or other insects creeping, crawling, or nibbling on the body, especially the genitalia, perianal area, or nipples."[10] This is not exactly the type of thing you'd want to tell someone about on a first date.

Lactophilia involves an attraction to breastfeeding, while *plushophilia* describes a sexual fixation on stuffed animals. *Mucophiles* find mucus to be especially alluring, and *eproctophiles* are enticed by, well, I'll just be blunt here: farts. And if that wasn't disagreeable enough for you, *hematigolagnia* involves becoming aroused by bloody menstrual pads.[11] The list goes on and on—truly—and it is curiously incredible for those of us who do not experience the allure of a taboo object of our sexual desire.

Unfortunately, some paraphilias involve non-consenting parties, which can push them from the realm of the bizarre but harmless into the invasive and potentially criminal. These include *pedophilia*, which is, of course, a disorder where one's sexual interests are focused on prepubescent children—and probably the paraphilia most likely to evoke moral outrage. But there are plenty of other paraphilias that can lead to illicit—and heinous—behavior. *Zoophilia*, for example, involves a sexual preoccupation with animals; it can sometimes be taken to the extreme of bestiality, or sexual activity between people and nonhuman animals (think back to David from Chapter 2 and his sexual dalliances with cats). And *necrophilia*, or a sexual attraction to corpses, is rare but not just an urban legend. There are, indeed, necrophiliacs in the world today (how many is unclear, as it is not something most people are inclined to admit to). Some only fantasize about sexual activity with a corpse, but others act on their fantasies. Searching a popular medical database for reports of necrophilia returns over 20 publications; reading through those case descriptions is much more macabre than any Stephen King novel, and the details are too grisly for me to write about here. But if you're the type of person who is titillated by things on the more gruesome end of the spectrum, then feel free to track down the sources linked to the endnotes at the end of this sentence.[12, 13]

Brain changes and paraphilias

In the scientific community, researchers typically distinguish between paraphilic *interests*, which occur when someone has an unusual sexual desire, and *paraphilias*, which occur when that unusual desire needs to be fulfilled for the individual to feel sexually satisfied. When the paraphilia begins to cause substantial distress, or when it causes harm to others, it can be classified as a *paraphilic disorder*.

Paraphilic interests are quite common. A survey of more than 1,000 men and women found that close to half (45.6 percent) of them admitted to a paraphilic interest, such as voyeurism, fetishism, or exhibitionism.[14] When it comes to actually engaging in paraphilic behavior, however, the prevalence falls sharply. In the same survey, only about one-third of the participants had taken part in the behavior they were interested in.

There is a lack of good studies on the prevalence of paraphilic interests, and you can probably guess why: these are just not things that many people are eager to openly talk about. For the same reason, we have even less reliable evidence on the occurrence of paraphilic disorders. We can be confident they are rarer than paraphilic interests, but it's difficult to put a number on it.

Additionally, talking about the neurobiological basis for paraphilia can be a precarious business, as in doing so you're bound to face some criticism—for multiple reasons. On one hand, there is a valid argument that attempting to identify aberrant brain activity underlying paraphilias is helping to perpetuate the idea that all such interests are pathological, instead of accepting that some of them may represent normal aspects of human sexuality. This viewpoint exists because of a long history of the medical establishment condemning any sort of sexual behavior that is not heterosexual and "normal" according to what many would argue are outdated, puritanical values. Indeed, many psychiatric professionals considered homosexuality to be pathological well into the 1970s, and it wasn't until 1973 that the American Psychiatric Association (APA) removed it from its list of disorders.[15]

Thus, one argument holds that groups like the APA still focus too much on sex that can lead to reproduction as being the only type of healthy sex. Along these lines, some assert that sexual interests and acts that don't harm others should never be considered deviant (while harmful and non-consensual sexual acts should still, of course, be deemed pathological). And according to this perspective, discussion that tends to associate non-harmful paraphilias with pathology is moving in the wrong direction.

I don't disagree with this take, but it's difficult to ignore that some paraphilic interests are quite rare. Thus, understanding what makes the brain work differently in these cases does (to me) have some value. It's important to stress, however, that a difference in brain function does not equate to a pathology, and atypical sexual interests can be perfectly healthy expressions of sexuality—as long as they are consensual and not harmful to others.

Discussions of the neuroscience of paraphilias are also fraught because some argue that attributing sexual behavior on the harmful end of the spectrum (such as pedophilia) to a neurobiological abnormality may imply that those who act on such urges are not responsible for their crimes. For example, if a pedophile can argue that pedophilic behavior was attributable to a brain tumor, are they free of responsibility for their actions? While this question deserves a deeper philosophical discussion, I would argue that understanding how a pedophile's brain might increase the likelihood of pedophilic behavior does not exonerate a pedophile from guilt if they engage in criminal acts. Even if someone has pedophilic interests due to aberrant brain activity, there are countless decisions they must make before acting on their interests, and those decisions do—according to how we view individual accountability in the world today—involve some element of moral responsibility.

That all being said, there are clear links between paraphilias and brain activity. We can look to one of the most common neurological disorders, Parkinson's disease, for a plethora of examples. Parkinson's disease is a condition characterized by the deterioration and death of neurons—a process known as *neurodegeneration*. The neuronal death in Parkinson's disease prominently affects areas of the brain that are involved in movement, which leads to the characteristic symptoms of the disease: tremor, slow and exceedingly effortful movements, rigidity, and postural impairments.

The neurotransmitter (a term that refers to chemicals neurons use to communicate with one another) dopamine is especially depleted in Parkinson's disease, as areas of the brain that produce dopamine are decimated by the neurodegeneration characteristic of the disorder. As dopamine levels fall, the symptoms of Parkinson's disease worsen, drawing a strong connection between dopamine depletion and the movement-related problems of the condition. Thus, doctors frequently prescribe Parkinson's disease patients drugs designed to increase dopamine levels, which can temporarily improve symptoms.*

These dopamine-increasing drugs, however, sometimes cause strange side effects that are linked to enhanced dopamine activity. The issues arise because dopamine plays an important role in motivation and pleasure-seeking. The details of that role are still a bit unclear, but the gist is that dopamine is heavily involved in enhancing motivation and driving us to pursue something our brains have identified as pleasurable. Of course, this means that dopamine is also a critical factor in addiction and other impulse control disorders.

Parkinson's disease patients who are taking dopamine-increasing drugs sometimes begin to feel euphoric highs—especially after an increase in dose. They also may also exhibit unusual and uncharacteristic behavior, such as pathological gambling, binge eating, compulsive shopping, and—most germane to our discussion—an abnormal preoccupation with sex. These types of side effects characterize what is known as *dopamine dysregulation syndrome,* or DDS.

---

* I use the qualifier "temporarily" because these treatments do not offer a permanent solution. They can reduce the severity of symptoms—sometimes for years—but at some point the drugs begin to lose their effectiveness as the disease progresses. When the effectiveness of the medication wanes, doctors typically increase the dose, but this leads to side effects. Those side effects eventually become comparable to the symptoms of the disease itself in terms of their disruptiveness to the patient's life, causing the drugs to no longer be a viable option for treatment.

One case of DDS, for example, involved a 74-year-old patient named Jim who had been dealing with Parkinson's disease for two decades. Over that time, Jim had also been taking dopamine-increasing drugs to control his symptoms, and he hadn't displayed any unusual behavior. When his doctor substantially increased the dose of his dopamine-enhancing drugs, however, Jim developed some startling new sexual tendencies.

Jim became completely preoccupied with sex. He got frequent erections and didn't try to hide them from friends and family. He wanted to have sex with his wife multiple times a day and became angry when she refused. Then, he propositioned his 15-year-old granddaughter while she was visiting. All of this was horrifying enough to his family, but to top it all off his wife walked in on him attempting to have sex with the family dog.[16]

While most patients taking dopamine-increasing drugs do not experience DDS, a small number do—especially after increasing their dose or first starting treatment. The many unusual sexual behaviors that have resulted include (but are not limited to): exhibitionism, sadomasochism, zoophilia, pedophilia, and fetishism.[17]

But Parkinson's disease patients aren't the only ones who can experience extreme changes in sexual interest after some disruption to brain chemistry or function. Take the case of Jacob, a 40-year-old schoolteacher who was in a stable marriage for two years when his life began to unravel precipitously. Jacob admits to always having an interest in pornography, but he claims that otherwise his sexual interests were typical (an assertion supported by those who know him).

A couple of years into his marriage, however, Jacob became fixated on sex. He began soliciting prostitutes for the first time, and his interest in pornography developed into something of an obsession. Unfortunately, Jacob's pornographic tastes expanded to include children.

Jacob built a collection of child pornography, then began to act on his new interest by making sexual advances toward his prepubescent stepdaughter. Jacob's stepdaughter told his wife about his behavior, which led to the discovery of Jacob's child pornography collection. Jacob was arrested for child pornography and molestation and removed from his home.

The night before his prison sentencing, Jacob began to experience a headache so severe that it sent him to the hospital. While there, he also complained of difficulties maintaining his balance, which prompted doctors to do a neurological examination. During that exam, Jacob repeatedly asked for sexual favors from female members of the medical staff, which doctors noted as unusual "hypersexual" behavior.

Doctors ordered an MRI of Jacob's brain and found a large tumor in his prefrontal cortex. They determined they could safely remove the tumor, so Jacob was scheduled for surgery.

After the surgical excision of the tumor, Jacob's sexual interests suddenly reverted to normal. In a surprising display of leniency, he was able to complete a Sexaholics Anonymous program in lieu of prison. Seven months after his surgery, he was deemed to no longer be a threat to his stepdaughter and given permission to move back into his home.

All's well that ends well, right? Except there was a slight hiccup. About a year after his surgery, Jacob developed a persistent headache, which was accompanied by an intense drive to collect pornography again. These symptoms led Jacob back to his neurologist, who again used MRI to determine that Jacob's tumor had begun to grow back. Another surgery removed the remnants of the tumor, and with that second surgery Jacob's paraphilic urges once more disappeared.[18]

Jacob's case is not an isolated one. There are an abundance of examples in the scientific literature of individuals who developed new sexual proclivities after a brain tumor, trauma, or disease. Sometimes their new sexual interests seem incompatible with the person they were before.

Of course, you could argue that these patients might have always had secret urges that their brain damage made it impossible to hide any longer. Perhaps, for example, Jacob's tumor interfered with brain circuitry that enabled him to suppress desires he knew were socially and morally abhorrent. For many, this explanation is easier to swallow than the idea that our core moral beliefs could disintegrate with some disruptions to our brain chemistry. On the other hand, the number of patients who previously displayed no interest in taboo sexual desires—but became obsessed with them after an unexpected influence on their brain—supports the argument that at least some of these cases involve completely new behaviors brought on by neurobiological changes.

We tend to consider things like our sexual orientation or romantic interests as fundamental aspects of our identity. It's uncomfortable to think that a slight manipulation of brain tissue or chemistry could dramatically change such an integral aspect of our personalities. And yet, as the cases described in this chapter demonstrate, we are one episode of brain damage away from loving someone—or something—we never thought possible.

6

PERSONALITY

One fall day in 1993, Dr. Richard Baer opened his mail to find the strangest letter he had ever received in his life.

Baer had been treating a patient named Karen, who had originally come to him severely depressed and contemplating suicide. Throughout their therapy sessions, Karen alluded to a childhood history of abuse, but it wasn't until she had been working with Dr. Baer for four years that Karen felt comfortable enough to provide him with all the details of her traumatic experiences. She claimed her father and grandfather had drugged and sexually assaulted her, let friends take advantage of her for money, and forced her involvement in ritualistic sexual practices.

Baer was still trying to process this shocking information when he received the letter. It said (the misspellings below are part of the original letter):

Dear Doctor Bear,

My name is Claire. I am 7 years old. I live inside Karen I listin to you all the time. I want to talk to you, but I don't now how. I play games with James and Sara. And I sing to. I don't want to die. Can you help me tie my shoes

Claire[1]

As strange as the letter was, some things started to make more sense to Dr. Baer after reading it. Early on in their therapy, Karen had divulged that she sometimes experienced periods of time that she had no memory of. One day, for example, she went grocery shopping but then blacked out. The next thing Karen remembered she was buying her son a hat in a department store. She never made it to the grocery store, but she couldn't recall why.

Karen admitted she had been having these episodes of lost time for most of her life. She had no recollection of long stretches of her childhood, and she couldn't remember ever having sex with her husband—even though they had two children together.

The letter thus confirmed a suspicion. Dr. Baer believed Karen had a condition commonly known as *multiple personality disorder* but officially referred to in the medical community as *dissociative identity disorder*. In dissociative identity disorder, or DID, the patient's behavior indicates the presence of two or more distinct personalities, sometimes called *alters*. When patients with DID "switch" personalities, they can present with a different name, gender, and voice. They may have unique mannerisms, and sometimes they even claim to have different physical characteristics, such as a need for eyeglasses.

Patients with DID tend to experience substantial memory lapses during the manifestation of an alternate personality. These memory deficits can be highly distressing, as the patient may have no recollection of what happened while one of their alters was in control. Thus, like Karen, DID patients can end up with prolonged episodes of time they're unable to account for.

After Dr. Baer received the letter from "Claire," other personalities began to emerge in Karen's therapy sessions. Eventually, Baer would meet 17 different personalities within Karen. They ranged from 2 to 34 years old and included males and females with distinct traits, interests, and physical attributes.

Claire, for example, was a 7-year-old girl who liked to play games and was scared of the dark. Katherine was a 34-year-old woman who loved classical music, opera, and playing the clarinet.

Holdon was a tall and masculine 34-year-old man who liked to bowl; he acted as Karen's protector.

Once Dr. Baer had met all of Karen's alters, he pursued the goal of integrating Karen's multiple personalities into one. This is a common treatment approach for DID patients. Instead of trying to get rid of a patient's alternate personalities, therapy often involves attempting to incorporate them into the patient's consciousness—to make a whole individual out of these separate parts, so to speak.

In Karen's case, integration of her personalities was finally accomplished in April 1998—nine years after she first came to see Dr. Baer. In 2006, Karen was able to discontinue therapy. Her symptoms of DID—along with the crushing depression that had brought her to see Dr. Baer in the first place—were in her past.

A failure to integrate

While the thought of having more than one personality residing inside your head may seem strange (or maybe not—more on that later), we are all familiar to some degree with the key feature of DID: dissociation (remember, DID stands for *dissociative* identity disorder). Typically, your brain is amazingly adept at integrating countless bits of information—about your perceptions, emotions, memories, identity, and so on—to give you a sense of continuity, both in terms of who you are and what is happening around you. Indeed, it does this so fluidly that it's sometimes difficult to recognize our experiences are pieced together from all these distinct components—unless dissociation happens. During dissociation, the brain fails to smoothly incorporate these different aspects of cognition, and conscious awareness can be interrupted.

Although it sounds dire, dissociation isn't necessarily debilitating. Mild dissociation happens frequently (even in healthy people) and can take the form of daydreams or short lapses in attention. When you realize, for example, that you've been looking out the window for several minutes at nothing in particular—only to catch yourself and remember that you should be writing a book—that is a mild form of dissociation.

In some cases, however, dissociation is more disruptive, causing a substantial disturbance in the brain's capability to maintain a continuous experience with the world. This can lead to symptoms such as depersonalization and derealization (recall these from Chapter 1). In patients with DID, dissociation is severe enough to impair the ability to maintain a stable identity, which may prompt the appearance of other seemingly distinct personality states.

Exorcisms, hypnosis, and DID

DID has gone by several different names over the years and was known as multiple personality disorder until 1994. The name was changed to DID in part to emphasize that DID patients struggle to integrate different aspects of their personality into a unified identity, rather than develop new and separate personalities that were never part of their core self (the latter is somewhat implied by the term *multiple personality disorder*). Regardless of the terminology, descriptions of patients who would be likely candidates for DID go back several hundred years.

The earliest cases of DID were typically interpreted as supernatural events. An account from the late 1500s, for example, describes a 25-year-old Dominican nun named Jeanne Fery who was thought to be possessed by the devil. Fery seemed to be inhabited by many personalities, some benign but others demonic (literally—they claimed to be demons). Her behavior would often shift quickly and dramatically, with sharp contrasts between her different personality states. At times she acted like a placid 4-year-old girl, but without warning she would switch into a devilish identity that exhibited extreme rage and aggression; once she even physically attacked an archbishop and his assistants. At other times she claimed to be Mary Magdalene.

The convulsions, contorted facial expressions, and episodes of headbanging, self-strangulation, and other disturbing behavior that are described in the written accounts of Fery's case sound like something out of *The Exorcist*. (It's possible, if not probable, that many cases of demonic possession throughout history may be better explained by DID.) Indeed, an exorcism is exactly how Fery's condition was treated—successfully, I should add, according to her church elders. Fery, however, was also under the solicitous care of her fellow nuns for a period of 21 months, which may have contributed more to her recovery than the exorcism.[2]

The idea that dissociation (rather than evil entities) could lead to the manifestation of another identity dates to the 1800s. At that time, the concept of dissociation was a fairly new one. In fact, just the idea that our mind could have both conscious and unconscious processes was novel, and the possibility that unconscious mechanisms could influence the way the mind worked intrigued psychologists of the day. These notions led to an interest in methods such as hypnosis, which had the potential to bring previously concealed subconscious thoughts to light—including the thoughts of alternate personalities, which were normally hidden beneath the surface in DID patients.

Hypnosis and other similar approaches were used to identify some of the first patients who would match the modern clinical definition of DID. One was a French man named Louis Vivet. At age 17, after a childhood rife with physical abuse, Vivet had a terrifying experience in which a viper wrapped itself around his left arm. Although the snake didn't bite him, Vivet was psychologically traumatized. That night he lost consciousness and experienced violent convulsions. He subsequently developed paralysis in his legs that didn't appear to be due to any physical cause (we'll discuss mysterious symptoms like this more in Chapter 7).

One year later, Vivet's paralysis suddenly disappeared—along with his memory for the entire prior year of his life. At the same time, he experienced a marked personality change: previously calm and polite, he became argumentative, impulsive, and borderline dangerous. Several months later, however, Vivet's legs again became paralyzed, and his calmer and gentler persona reemerged. This alternation between physically capable but quarrelsome, and paralyzed and mellow, continued for years—much of it occurring under the observation of interested physicians who (with the aid of hypnosis) recorded up to 10 distinct personality states in Vivet.[3]

Doctors identified many other cases of DID from the nineteenth century on, but awareness of the condition increased markedly in the second half of the twentieth century with a couple of well-publicized cases. One was Christine Sizemore, an American woman whose struggles with DID were depicted in a book (subsequently made into a popular movie) called *The Three Faces of Eve*. In the 1970s, another influential book about DID was published and eventually made into a TV movie. It was titled *Sybil*, and told the story of Shirley Mason, a woman who allegedly had 16 personalities. (Sybil was a pseudonym used to protect Mason's privacy, but her true identity was discovered after her death.)

These prominent DID cases piqued the public's curiosity, and soon writers were commonly using DID as a plot mechanism in many other books, movies, and television shows. The number of DID cases rose along with the increased awareness of the disorder. By some estimates, there were only 50 known cases of DID before *Sybil* was published, but more than 20,000 had been diagnosed by 1990.[4]

Sybil and Shirley Mason's case, however, also exemplify some of the controversy surrounding DID. A couple of decades after the publication of *Sybil*, evidence emerged that the psychologist who treated Mason, Dr. Cornelia Wilbur, may have used therapeutic methods that encouraged Mason to assert the presence of multiple identities—even though their existence was purely imagined.[5] Thus, some have suggested that Wilbur's approach to treatment—and perhaps her ethics—may have been influenced by her desire to profit from the case both professionally and monetarily.[6]

Shirley Mason's case and others like it have led some to doubt the validity of DID diagnoses. According to these skeptics, the increased number of DID cases we've seen in the last 50 years is partly due to therapeutic practices that (either intentionally or unintentionally) prompt a patient to believe they are inhabited by multiple personalities. Additionally, awareness of highly publicized cases might cause patients (again, either intentionally or unintentionally) to try to emulate symptoms of DID. From this perspective, DID is a learned behavior—not a psychiatric disorder.

The available evidence, however, suggests DID is a valid diagnosis. Studies have consistently found differences in psychological characteristics between patients with genuine DID diagnoses and study volunteers who are asked to pretend to have DID symptoms.[7] Moreover, tools used to diagnose DID tend to do so reliably.[8] And researchers have also identified measurable differences in the way the body and the brain work when patients with DID are in an alternate personality state. Because some of these variations in biological functioning seem impossible to fake, they provide perhaps the most convincing piece of evidence in support of the legitimacy of the disorder.

The biological basis of DID

One startling case involved a patient who was by all indications blind when she began treatment for DID. After four years of treatment, however, psychiatrists discovered that one of her alters (a young male) had normal vision. It seems, of course, that this patient's blindness was not due to some physical defect in the eye or visual system, but instead attributable to psychological factors (this type of blindness is sometimes called *psychogenic blindness*). Nevertheless, a university eye clinic had verified the patient's blindness to support her claim for disability payments, and it appeared to be a genuine handicap. With further therapy, the patient regained vision in some personality states but remained blind in others.[9]

Studies have also found differences in brain function in people with DID. One experiment, for example, looked at brain activity in a small group of DID patients while they were in distinct personality states. The researchers hypothesized that, if DID is a disorder that truly involves switching into alternate personalities, then there should be unique patterns in brain activity associated with those identities. And that's just what the results suggested: the different personality states were linked to characteristic patterns of brain activity, suggesting that a switch into an alternate personality is associated with changes in the way the brain is working.[10]

Still, scientists have struggled to explain what happens in the brain to cause DID. One popular hypothesis, however, focuses on the brain's response to trauma such as physical or sexual abuse in childhood. Most DID patients have experienced some sort of trauma,[11] and many scientists believe that people with DID develop alternate selves to cope with overwhelming feelings brought up by memories of traumatic events. This may lead to the compartmentalization of these powerful feelings in an alternate identity, which—in a sense—protects the host from the associated emotional pain. This explanation for the origin of DID is known as the *trauma model* of DID.

Understanding that most DID patients have experienced trauma, it's probably not a surprise that many individuals with DID also suffer from post-traumatic stress disorder, or PTSD—a condition that involves the frequent reexperiencing of a traumatic event through flashbacks, nightmares, and the like. And there may also be neurobiological links between PTSD and DID; in both disorders we see abnormalities in brain structures that are involved in processing traumatic experiences.

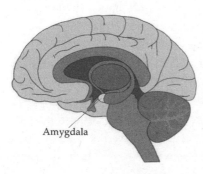

Amygdala

One such abnormality involves the *amygdala,* a small group of neurons in the temporal lobe. The word *amygdala* is Greek for "almond," and the structure got its name because it has something of an almond-like shape. Although we tend to refer to the amygdala in the singular,* there are actually two of them—one on each side of the brain.

Despite the relatively unassuming size of the amygdala, it makes complex contributions to our emotional experiences. The amygdala plays a fundamental role in orchestrating emotional responses, and it helps to form the types of enduring memories that are typically connected to emotionally poignant events. But an especially large amount of evidence links the amygdala with a particular emotional response: fear.

Many studies, in both animals and people, suggest the amygdala plays an important role in fear. When we encounter something in our environment that poses a threat to us, neurons in the amygdala become highly active. They send signals to other parts of the brain to initiate what's known as a *fight-or-flight* response. You've probably heard of the fight-or-flight response before, if nowhere else than in your high school biology classroom. But even if the term is new to you—or if you've forgotten it like I did almost everything I learned in high school biology—the concept isn't foreign because you've experienced a fight-or-flight response countless times.

A fight-or-flight response causes a well-known constellation of physical effects that occur whenever you're afraid or extremely nervous: your heart rate increases, your blood pressure goes up, your breathing quickens, your pupils dilate, etc. These physical changes have a purpose: to prepare your body to either run or fight by doing things such as increasing the amount of oxygen-rich blood being sent to your muscles to get them ready to act, dilating your pupils so you can see more of what's around you, and so

---

* The tendency to refer to multiple brain structures in the singular is common in neuroscience. Most of our brain structures are paired, meaning there are two of them (one on each side of the brain), but very often this isn't reflected in how we write or talk about such structures. Even though this can be a somewhat confusing convention, I'll follow it rather than use the lesser-known plural forms of these words (e.g., amygdalae), which I worry might be even more confusing.

on. At the same time, the response inhibits processes that aren't worth devoting energy to at the time, such as bladder contraction (because wetting your pants while you're fighting would just be unfortunate), digestion, etc.

Scientists think the fight-or-flight response helped our species to survive. Primitive humans would have encountered many instances where their lives hinged on the body's ability to get ready to fight or flee in short order (such as after encountering a predator). Without a fight-or-flight response, your ancient human ancestors might not have acted quickly enough to live to tell the tale of their run-in with a lion on the savannah, causing the lineage that led to your birth to have been cut short tens of thousands of years ago.

It's worth noting that the fight-or-flight response is a reaction both to situations that are physically dangerous and those that are psychologically threatening. So, whether you're running from a vicious dog you've encountered on your morning walk or giving an oral presentation to a group of coworkers, the characteristic signs of a fight-or-flight response are occurring; the degree, not the response itself, is what's different.

While the fight-or-flight reaction evolved as an important safety mechanism, overuse of the response can be detrimental. In addition to anxiety, it can cause the secretion of high levels of stress hormones that, when left to circulate too long in the bloodstream, have damaging effects ranging from impairing the integrity of blood vessels to causing the death of neurons.

Therefore, we need to be able to regulate the activity of the amygdala to keep it from overreacting to things that don't pose a legitimate threat to our well-being. In steps the prefrontal cortex. As I mentioned in Chapter 3, the prefrontal cortex is involved in rational thought, and in this situation

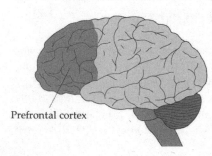

Prefrontal cortex

it is believed to play an important role in inhibiting unwanted knee-jerk reactions from the amygdala. The prefrontal cortex aids your brain in determining if something in the environment truly poses an immediate threat. If it doesn't, the prefrontal cortex can help to dampen the amygdala's response—acting in a sense like the logical voice to calm the amygdala's emotional reaction.

Imagine, for example, that you did get attacked by a vicious dog on your morning walk and suffered several minor bite wounds because of it. Then, you are visiting a friend when their good-tempered dog comes ambling toward you, looking for you to pet him. After your recent harrowing canine assault, seeing any dog approaching might generate a fearful reaction. But just as your amygdala's activation level starts to get very high, your prefrontal cortex can kick in and turn it down, reminding you that—despite your other negative experience—this current dog does not seem to pose a threat.

In some cases, however (such as in patients with PTSD) the inhibitory link between the prefrontal cortex and amygdala does not function the way it should, and the amygdala's activity goes unchecked. This might cause the amygdala to react to stimuli related to a traumatic event—or even just to memories of the event—as strongly as it reacted to the traumatic event itself. It's almost as if the event was happening all over again, and the amygdala decides it warrants a full-fledged fight-or-flight response. This may cause patients with PTSD to feel like they are continually reliving a traumatic experience.

To explain how some patients who have undergone trauma develop DID, researchers have proposed that the inhibitory pathways from the prefrontal cortex to the amygdala can sometimes be *overactive*. This overactivity occurs in response to trauma and causes excessive inhibition of the amygdala, perhaps in an attempt to dull the overwhelming emotional response to distressing memories. The increased inhibition is associated with emotional detachment and dissociation.[12] Thus, the excessive amygdala inhibition might act as a self-defense mechanism to dull emotional pain—but the self-defense mechanism works too well, leading to such extreme emotional inhibition that the patient experiences dissociation and, in rare cases, a fracturing of their identity.

This hypothesis is far from being the last word on what happens in the brain to cause DID. Other brain structures and mechanisms have also been implicated, and there is much more research that needs to be done before we have a clear idea of the neurobiological origins of the disorder. At the same time, the trauma model of DID is a viable starting point. It might explain features of DID in some patients, and it could pave the way for a better overall understanding of the disorder in the future.

The different flavors of dissociation

Many researchers are also looking further into the experience of dissociation itself, as a better understanding of this unusual psychological state might help us to learn more about other disorders that are characterized by severe dissociation. There are several such disorders besides DID, and their symptoms can be just as disruptive as the presence of multiple personalities. Like DID, they are all often linked to a preceding traumatic—or highly stressful—event.

In *depersonalization/derealization disorder*, or DPDR, someone suffers from episodes where they suddenly feel detached from the world or get the unshakable sense that the world around them is not real. They might feel like they are living in a dream or are a distant observer of their own lives. These feelings can get so extreme that someone might have the sensation they are seeing themselves from outside their own body.

One patient with DPDR described how he was walking along a city street when he suddenly felt like he was looking down on himself from the awning of a nearby store. That unexpected episode was the first of many; indeed, it started what became a lifelong struggle with DPDR. During the next 20 years, he had repeated out-of-body experiences. "Since then," he said, "I've...never completely felt like I was back in my body."[13]

Dissociative amnesia involves a severe disruption in memory that can last anywhere from minutes to years. The memory loss might be general and widespread, or selective only for specific events and details. In rare cases, it's accompanied by what's known as a *fugue state*, where the patient, lacking the memories that would maintain an attachment to their current life, travels away from home without warning. Sometimes they may wander aimlessly, and other times they may act with a distinct purpose that's known only to them and has no logical connection to their pre-amnesic life. At its most extreme, someone who experiences dissociative amnesia may move to a different area and assume a new identity—all without any awareness of their past life. For patients who go to these lengths, recovering memories of their previous life can be a traumatic experience in and of itself—as it can involve a shocking revelation that they have not always been the person they now think they are.

Extreme stress or a traumatic event often precipitates dissociative amnesia, but unlike other more common types of memory disturbances, dissociative amnesia frequently affects people under the age of 40.[14] Take, for example, the case of Logan, a 20-year-old man who had no previous medical or psychiatric problems before his mother brought him to the hospital one Monday morning. He had been fine just two days prior, but on Sunday he showed up at his job confused. He didn't recognize his coworkers, and he asked his supervisor what he was supposed to be doing at work. He was sent home right away, but he failed to recognize his mother, siblings, or his dog when he got there.

Logan's mother encouraged him to sleep, hoping some rest would restore his mental health. Despite his confusion, Logan complied with his mother's wishes. He went to sleep for several hours and woke up in the afternoon, but then he left home without telling anybody. His mother noticed that Logan was gone and began frantically sending him text messages, but Logan didn't respond. She called a few of his friends and organized a small search party. Eventually, they found Logan in the parking lot of a convenience store. He didn't remember how he got there, or why he had come in the first place.

Logan's amnesia had been preceded by a particularly difficult breakup a week prior. After being diagnosed with dissociative amnesia in the hospital, Logan was discharged and began therapy to attempt to retrieve his lost memories. His amnesia persisted for over three months, but eventually he started to regain his memory and was able to return to his job and previous life.[15]

While the neurobiology underlying other dissociative disorders such as DPDR and dissociative amnesia is still not completely clear, neuroscientists hypothesize the involvement of similar mechanisms to those believed to occur in DID. In DPDR, for example, one proposition is that the prefrontal cortex exerts an excessive inhibitory influence over emotional regions of the brain (such as the amygdala) in an attempt to manage painful emotions linked to trauma. Instead of this mechanism leading to a fracturing of one's identity, however, it causes a deficit of emotion that can generate the feelings of detachment characteristic of the disorder.[16]

The prefrontal cortex might play a similarly important role in dissociative amnesia. According to one hypothesis, the prefrontal cortex and other regions may attempt to withhold distressing trauma-related memories from conscious awareness as a way of protecting the individual from experiencing intense emotional pain. The effort required to do this, however, might interfere with mechanisms needed to maintain one's memories. These memory mechanisms may be further disturbed by the release of large amounts of stress hormones, which can have a memory-disrupting effect.[17]

Although these other dissociative disorders clearly can have a substantial impact on someone's life, DID is still the best-known of the dissociative disorders. This is likely due to its incredibly unique symptoms. As I searched the scientific literature for cases of DID that struck me as especially noteworthy, most that I found resembled Karen's: an individual with DID and numerous alters. One case that I stumbled upon, however, stood out from the rest in terms of its outlandishness.

A thirst for blood

Abdul was a 23-year-old man who came to the hospital complaining of a strange addiction. His obsession, he claimed, began after a series of traumatic events: the witnessing of his uncle's murder, watching his friend commit a violent homicide involving decapitation and genital mutilation, and the death of his 4-month-old daughter.

After his daughter's death, Abdul developed an intense desire that was straight out of a horror movie: he had an incessant thirst for human blood. Initially, he satisfied this craving by cutting himself with razor blades and collecting enough blood in a cup to drink. But eventually he began to crave the blood of other people as well. He felt an intense yearning to drink others' blood that he said was "as urgent as breathing."[18]

Unfortunately, Abdul acted on his urges. He was arrested multiple times for stabbing or biting people and attempting to drink their blood. Abdul's father often got blood from blood banks to appease his son and reduce the likelihood of these violent attacks.

Abdul would typically have no recollection of his vicious assaults; he reported many episodes of losing time in general. He claimed he would often find himself in a new place with no idea how he got there, and he said he frequently ran into people on the street who called him by different names. Abdul eventually ended up in the hospital, where he was diagnosed with DID, major depressive disorder, and PTSD. It's likely the different names people referred to him by were the names of alters who had been in control during Abdul's episodes of lost time.

It's important to note that DID does not typically involve violent behavior. Abdul, however, was experiencing significant delusions and hallucinations, which are less common symptoms of DID. He claimed to frequently see a "tall man with a black coat" nearby, as well as a young child who urged him on to violent acts. The child would say things like "jump on him" or "choke him," and Abdul would comply.

Abdul stayed in the hospital for two weeks; during that time, doctors prescribed several medications to treat his depression and delusional thinking. After Abdul was discharged, he got in a violent argument with his wife's family and was admitted to the hospital again, this time for three weeks. After that second hospital stay, Abdul stated that he no longer felt the need to drink blood, but he continued to have episodes of losing time. At his last follow-up, Abdul was pessimistic, stating, "This mess can end by my death only."

Cases like Abdul's, where there is some violent behavior linked to an alter, have captivated the public's interest since the first cases of DID were recognized. Perhaps it is due to the unease caused by feeling that another personality inside of us has the potential to emerge and commit some heinous act. Or maybe it's because we see this type of behavior as an expression of pent-up urges we all experience.

But I also believe part of the fascination with DID is that we all—at some level—recognize our personalities are not consistent. As much as we would like to think of our psyche as stable and reliable, in truth it is made up of a jumble of thoughts that sometimes seem like they come from a multitude of inner selves.

For me, at least, it often makes more sense to think of myself as consisting of many different—albeit tightly interconnected—personalities. Sometimes I'm social, sometimes I'm reclusive, sometimes I'm a leader, sometimes I'd rather follow. The difference between patients with DID and me, however, is that my personalities are all integrated with one another—and aware of each other's existence. When I display a tendency in my patterns of thought, I don't dissociate, but incorporate that pattern into my overall conceptualization of who I am.

Nevertheless, we tend toward thinking of ourselves in dichotomies. We are either social *or* shy, leaders *or* followers, and so on. But instead of thinking of ourselves as either/or, perhaps we are better thought of as a combination of all—a mishmash of tendencies that alternate in their frequency of presentation.

And from that perspective, maybe DID is more representative of the spectrum of normal human personality than it might seem at first. Understanding, then, what happens in the brain to cause the failure to integrate personalities that is characteristic of DID may be informative not just about DID, but about the human experience itself.

7

BELIEF

In 1973, Dr. Clifton Meador met a new patient named Sam who was dying of esophageal cancer, but their introduction didn't go well. Sam hid under the covers of his hospital bed and refused to come out. "Go away," Sam said feebly, but with evident irritation. "Leave me alone."[1]

When Meador pulled back the covers to look at Sam, he saw an old, unshaven man who could barely open his eyes and appeared "nearly dead." Little did Meador know that this frail elderly man would radically alter his perspective on how belief can influence health, life, and death.

Meador talked to Sam's wife, Sarah, and learned that several months ago Sam had been given only months to live. Sam's diagnosis was stage 4 esophageal cancer, and the cancer had spread to the liver (a typically incurable situation). Sam and his wife, both in their 70s, had moved to Tennessee to be closer to family who could help with Sam's end-of-life care; they met Dr. Meador in a hospital in Nashville.

After a few days of supportive treatment, Sam regained some strength and emerged from what had been a debilitating depressive state. He was able to get out of bed and walk up and down the hall several times a day, and soon he gained a few pounds. Eventually, Sam began to open up to Dr. Meador. Sam told his new doctor a heartbreaking story about how his life had changed in the past two years.

Sam explained that he had married Sarah only a few months ago; she was his second wife. A year and a half prior, Sam had been living with his first wife, June, whom Sam described as his soul mate. Sam and June had loved boating, and after years of saving they were able to buy a retirement home on a large lake. Sam had planned to live out the rest of his days in their new lake house.

Then, one night while Sam slept, a natural disaster decimated his life plans. An earthen dam nearby burst, causing a flash flood that slammed into their house, sweeping it—and Sam and June— into a nearby river. June's body was never found. Sam narrowly escaped death by clinging to the wreckage of his house. As Sam relayed this story to Meador, he sobbed and said, "I lost everything I ever cared for. My heart and soul were lost in that flood that night."

Less than six months later, Sam began to have severe difficulty swallowing. While investigating the cause of his swallowing problems, doctors discovered cancer in Sam's esophagus. During surgery to remove the cancer, it became clear the malignancy had spread to Sam's stomach.

Esophageal cancer unfortunately carries a grim prognosis— especially once it has spread beyond the esophagus. On average, less than 6 percent of such patients live for five or more years after their diagnosis.[2] Despite this bleak outlook, Sam met his second wife, Sarah, after his cancer diagnosis, and she vowed to support him through his battle with the disease.

After hearing Sam's heart-wrenching story, Meador asked him, "What do you want me to do?" The implication of the question was clear. Sam's time was coming soon, and Meador had a couple of options: he could provide palliative care to make Sam's death a less agonizing one, or he could try to help Sam prolong his life as much as possible. The latter option, of course, might mean accepting substantial pain and discomfort as a trade-off for longevity.

"I'd like to live through Christmas, so I can be with my wife and her family," Sam said. "Just help me make it through Christmas. That's all I want." It was October.

Meador created a treatment plan for Sam that he could follow at home, and Sam left the hospital near the end of October. At that time, Sam outwardly appeared to be in good health. Meador attributed this to exceptional nursing care and mentioned that— had he not known Sam's diagnosis—he would have expected a good outcome. Meador continued to see Sam regularly from late October through Christmas. During that time, Sam was in generally good spirits and health, all things considered.

Just after New Year's, however, Sam came back to the hospital. It was clear he had deteriorated considerably, and once again he looked very sick. Sam, however, had just a slight fever and no especially dire symptoms—nothing that would indicate impending death. But he told Sarah that, now that he had lived through Christmas, he was going to the hospital to die. He died in his sleep 24 hours later.

After Sam's death, Meador performed a routine autopsy (routine autopsies were much more common in the 1970s than they are today), and what he found shocked him. He had expected to see that Sam's esophagus and stomach—and probably other organs as well—were riddled with cancer. But Sam's esophagus was cancer-free. He did have a small cancerous lump in his liver, but it was not large enough to interfere with Sam's liver function; Sam likely would have lived with it for years before it began to produce any symptoms. There was no apparent cancer anywhere else in his body. Meador found bronchial pneumonia in a small region of Sam's lung, but it was not enough to kill.

Thus, Sam had been living under the burden of a fictitious disease—a misdiagnosis. He wasn't exactly healthy, but he also did not have terminal cancer. Meador couldn't even determine a cause of death. Sam died *with* pneumonia and cancer, but he did not die *of* either disease.

After much deliberation, Meador decided that Sam had died because Sam thought he was going to die. Meador hypothesized that the accuracy of Sam's belief was overestimated because his doctors (authority figures he trusted) had endorsed it. And everyone around him had echoed that same belief, to the point where his brain—certain of the reality of the situation—convinced his body of it as well.

While this perspective may not align well with a Western medical viewpoint, there are a number of similar examples of deaths that seem to be strongly linked to the belief that one was expected to die. Another case, for example, involved a 22-year-old patient whom I'll call Gabrielle. Gabrielle went to the hospital after experiencing shortness of breath, chest pain, dizziness, and fainting that had been going on for six weeks. When a doctor interviewed her, Gabrielle was in a panic.[3]

Gabrielle explained that she had been born on Friday the 13th. The midwife who delivered her had also delivered two other children on the same day. Afterward, the midwife told Gabrielle's mother that all three children were cursed (not the type of behavior that would earn a midwife high ratings on Yelp). The first child, she said, would die before she turned 16. The second was destined to die by the age of 21. Gabrielle, the midwife declared, would perish before she turned 23.

Ominously, the first-born of the three died in a car accident on the day before her 16th birthday. The second, who was very aware of the curse and quite apprehensive about it, made it to her 21st birthday and decided to go out to celebrate her evasion of the hex. As she walked into a bar that night, she was hit by a stray bullet and died. Gabrielle's 23rd birthday was rapidly approaching, and she was terrified.

Gabrielle began experiencing episodes of hyperventilation that became increasingly more frequent and intense the closer she got to her birthday. On the day before her birthday, she started wheezing and sweating profusely—symptoms that were precipitated and worsened by her state of extreme fear. She was dead soon after.

An autopsy showed that Gabrielle had pulmonary hypertension (high blood pressure in the arteries that carry blood to the lungs) which led to problems with normal heart function. In short, there were factors that could have caused her death. Nevertheless, doctors were uncertain if it was these factors or her intense fear of the curse that truly led to Gabrielle's demise.[4]

Voodoo death

Death that seems inextricably linked to strong belief goes by several names. The more scientific term is *psychogenic death*, which refers to the potential role psychological state and strong emotions may play in these cases. But the more common terms, *voodoo* or *hex death*, imply that mysterious forces are at play.

Of course, most scientists aren't satisfied with attributing death to the black arts, so they've looked for other, more rational interpretations. Some of the earliest attempts at scientifically explaining psychogenic death date to the 1940s and are attributed to Walter Cannon, one of the most influential of all American physiologists. Whether you realize it or not, you're at least vaguely familiar with some of Cannon's work just from reading the last chapter. He coined the term "fight-or-flight" to describe the way the nervous system responds to threatening events, and he was the first to work out some of the biological mechanisms behind the fight-or-flight response.

While we talked in Chapter 6 about the amygdala's role in initiating a fight-or-flight reaction, the execution of the response itself involves a part of the nervous system known as the *sympathetic nervous system*. The sympathetic nervous system is a collection of nerves that extend throughout the body; they can energize your body for immediate action as well as communicate with the brain to orchestrate the release of hormones needed to prepare your body for a longer lasting response to a potential threat. Cannon suggested psychogenic death might be due to overstimulation of the sympathetic nervous system, which leads to the release of large quantities of the hormone adrenaline and corresponding physical effects that prompt the onset of shock—and, in some cases, death. Thus, according to Cannon's view, the victims of psychogenic death were, quite literally, scared to death.[5]

But Cannon's hypothesis explains sudden death, while often psychogenic death can take weeks or longer to occur (as we saw in Sam's case). Thus, Cannon's hypothesis doesn't demystify all cases of psychogenic death, and other attempts at explaining the phenomenon in purely physiological terms have similarly fallen short. This has led some scientists to grudgingly accept that there might be an influence of anticipation in cases of psychogenic death. In other words, in some of these patients, death seems to be occurring—at least in part—because they expect and believe that it will.

The power of belief

The idea that belief can affect health is certainly not a new one. In the 1800s, for example, doctors frequently used placebos in their medical practice to alleviate patients' symptoms. Placebos are inactive substances disguised as medicine; they're given primarily to create the impression the patient has just received something that will help their ailment. In the 1800s, commonly used placebos included bread pills, sugar pills, water injected under the skin, and colored water to drink—all given to patients with the assurance they were getting a formula that was certain to expedite their recovery.

Doctors in the nineteenth century didn't use placebos to trick or swindle their patients; they used them because they worked. At a time when true treatment options were somewhat limited, without placebos doctors often didn't have much else to offer besides well-intentioned advice. But they knew providing something tangible that the patient believed would make them feel better was more likely to lead to some improvement than giving the patient nothing at all. According to some estimates, in that age physicians used placebos more frequently than all other medicines combined,[6] a practice that continued until the first half of the twentieth century, when ethical qualms about deceiving patients became more pervasive.

Ironically, it wasn't until placebos became less common in medicine that scientists began to learn just how powerful they really could be. The beginnings of modern research on placebos and their effects (known as *placebo effects*) can be traced back to a surgeon named Henry Beecher. Beecher had become interested in the placebo effect while working in a battlefield hospital during World War II. He recounted how at times he was faced with shortages of potent analgesics such as morphine, and he would inject wounded soldiers who were desperate for pain relief with saline instead, telling them it was morphine to appease them. To Beecher's surprise, he found that the saline solution often worked almost as well as morphine in placating injured soldiers.[7] This suggested to Beecher that pain relief was, to some degree, influenced by his patients' expectations.

Intrigued by these observations, Beecher dove into research on placebos after the war. To learn more about the effectiveness of placebos, Beecher randomly selected a group of 15 studies that compared placebos to active drugs in the treatment of conditions ranging from seasickness to severe post-operative wound pain. Combined, the studies included just over 1,000 subjects, and Beecher found that a placebo alone alleviated the symptoms of more than one-third of them.[8]

Later researchers have suggested that Beecher's estimate was a bit exaggerated, although not intentionally. Beecher simply did not account for other factors that could have played a role in the patients' improvement, such as the tendency for some symptoms to naturally fade over time (even if neither a placebo nor a drug was provided).

Regardless, Beecher's findings shook the world of clinical

medicine. They not only suggested that the placebo effect was more powerful than anyone had imagined, but they also implied that a substantial proportion of drug effectiveness itself was due to the placebo effect. In other words, if we experience a benefit every time we take a placebo pill just because we think it will make us feel better, then when we take a real drug some portion of its effect is also due to us thinking it will help (while the rest is attributable to the active ingredients in the drug).

Not all in your head

After Beecher's groundbreaking study, researchers became more interested in understanding what is happening in the brain and body to cause the placebo effect. At first, scientists thought the placebo effect was mostly mental; it was assumed that if patients believed a treatment could make them feel better, they would feel better primarily due to the power of positive thinking. As research techniques became more sensitive, however, we began to realize that placebos don't just change beliefs, they also influence the functioning of the body (and the brain).

One of the seminal studies to demonstrate that placebos could affect the way the body works focused on substances called *endorphins*. The name "endorphin" is a portmanteau of the words *endogenous* and *morphine*. It was coined to reflect that endorphins are natural bodily substances (the word *endogenous* refers to substances that are produced within the body) that have pain-relieving effects similar in some ways to those of morphine. Like morphine, endorphins act on the nervous system to inhibit pain signals before they reach the brain.

In the 1970s, scientists began to suspect that endorphins might play a role in placebo-induced pain relief. They hypothesized that placebos prompt the release of endorphins or similar substances, which could then block pain signals to cause analgesia. To test this hypothesis, one group of scientists went to a dental clinic and recruited 51 patients who were about to have their wisdom teeth removed. The investigators convinced these patients to take part in a study that would involve three groups: one group would receive morphine to manage their postsurgical pain, another group would only receive a placebo (no morphine), and the last group would receive a substance called naloxone, which blocks the actions of endorphins and might thus cause an *increase* in postsurgical pain. (How the researchers convinced the participants to be part of an experiment that could lead to increased dental pain is perhaps the greatest mystery of this study.)

The patients who received naloxone reported significantly more pain than those who received the placebo alone.[9] Since blocking the action of endorphins caused a decrease in the placebo effect, it suggested that the placebo effect in this case was—at least in part—mediated by the natural release of endorphins. This study, then, provided some of the first evidence of a physiological mechanism for placebo effects. Put another way, the study demonstrated that placebo effects (at least for pain) are not purely psychological. We don't feel less pain when we get a placebo just because we think we should; we feel less pain because some pain signals are actually prevented from reaching our brain.

Since this 1970s experiment, a slew of other studies have also concluded that placebo effects are often attributable to physiological mechanisms. These studies show that placebos can cause changes to brain activity, such as affecting the release of neurotransmitters[10] and hormones.[11] Placebos can also alter the functioning of the immune system,[12] the heart,[13] the gastrointestinal tract,[14] the respiratory system,[15] and more.

Through these influences, placebos can have an astounding range of effects on the functioning of the brain and body. For example, they can increase energy, improve athletic performance,[16] reduce stress,[17] help you sleep,[18] keep you awake,[19] decrease appetite[20]—the list goes on and on. Additionally, they have potential therapeutic benefits—in treating pain and a range of other conditions. Placebos, for instance, have a substantial impact on depression; according to some estimates, the placebo effect is responsible for up to 80 percent of the improvement seen in most patients taking antidepressants.[21] Placebos can also transiently improve symptoms in complex neurological disorders such as Parkinson's disease[22] and epilepsy.[23] Incredibly, placebo surgery— where an incision is made but no actual surgery is performed—has mitigated the symptoms of conditions such as osteoarthritis[24] and meniscus tears.[25]

Many scientists would agree that the placebo effect is one of the more surprising—and hard to explain—scientific phenomena. At the same time, it's quite a nuisance to researchers trying to decide if a "real" treatment works. Because patients may experience a placebo effect in response to any substance from which they might expect a benefit, we always need to subtract the placebo effect to determine the effectiveness of the real treatment. Put another way, to know how effective a drug is, we must understand how well it works above and beyond the placebo effect it is also likely to cause.

Harmful beliefs

Another important consideration when it comes to the influence expectation can have on treatment and health is the evil twin of the placebo effect: the *nocebo effect*. The word *nocebo* comes from the Latin *nocere*: to harm, and the nocebo effect occurs when negative expectations about a treatment lead to deleterious effects, even if the treatment alone doesn't truly cause them. The nocebo effect often occurs in clinical trials for drugs where participants aren't aware if they've received a placebo or the real treatment.*

---

* It is common practice in clinical trials to keep participants in the dark about whether they've received a drug or a placebo—an approach known as *blinding*. There are multiple reasons for using blinding in an experiment, but

Some patients who are unknowingly given a placebo report side effects that align with the side effects they expected to experience in response to the active drug.[26] Thus, similar to the placebo effect, patients who anticipate side effects are more likely to have them—whether they're taking a real drug or an inactive substance.

One study explored the nocebo effect in a group of patients taking finasteride (better known as Propecia) for benign enlarged prostate—a condition that affects a significant proportion of older men and leads to symptoms such as difficulty urinating, a frequent need to urinate, and an inability to completely empty the bladder (an especially uncomfortable trifecta of issues). Finasteride is linked to sexual side effects such as erectile dysfunction and loss of libido, but some researchers have speculated that the nocebo effect might influence the frequency of these side effects.

To test this hypothesis, scientists gave finasteride to 120 men to treat their prostate symptoms, but only informed half of them that sexual problems were possible side effects of the drug. Thus, half of the men believed they might have to endure some degree of sexual dysfunction, while the other half had no such expectation. The expectation proved to play a critical role in the occurrence of side effects—three times as many men who had been told about the possibility of sexual issues reported having them.[27]

Placebo and nocebo responses are thus attributable largely

---

one is to balance out the influence of the placebo effect. If blinding wasn't used, then participants who knew they were receiving a drug might experience a larger placebo effect due to the expectation of positive effects from the drug. Participants who knew they were not receiving a drug, on the other hand, would have little expectation of a benefit, and thus a diminished placebo effect. This discrepancy between the two groups would make it difficult to gauge the true effect of the drug, because the placebo effect would cause an exaggeration of the drug's effectiveness. By keeping expectation in both groups the same, researchers can assume a similar degree of placebo effect, which can then be subtracted to determine the overall effectiveness of the drug.

to our expectations: when we believe something will affect us in a certain way, it's more likely to have effects that align with our beliefs. As we've seen, however, these phenomena are not just in our minds. They're associated with distinct changes in how the brain and body work. Is it possible, then, that similar mechanisms of expectation could underlie psychogenic death? Unfortunately, the answer to this question is still unclear. But the evidence surrounding psychogenic death, placebos, and nocebos suggests there is an undeniable power of belief when it comes to the functioning of our bodies.

And this potential for belief to change how the body works underscores an important point: our mental and physical worlds are not independent of each other. We often emphasize the distinction between psychological and physical symptoms, as if our brains and the thought patterns they produce are somehow isolated from the rest of our corporeal selves. What neuroscientists now realize, however, is that the body and the brain are constantly influencing each other, and their dynamic relationship is tremendously important for our overall functioning. This understanding is helping scientists to develop a more complete picture of how people's brains and bodies interact—with each other and the world around them—and it's helping to treat conditions that have been historically misunderstood due to the assumption that the brain and the body operate as two distinct entities.

Withdrawn from the world

In August 2019, a 9-year-old Afghani girl named Mina witnessed a brutal stabbing in a refugee camp in Greece. Mina had already been through several very difficult years. She was injured in a bombing in Afghanistan in 2015, and her brother died in the same attack. She spent much of the next two years separated from her family while undergoing medical treatment (including multiple surgeries) to repair the damage to her leg suffered in the explosion. She still had difficulty walking and had to use a wheelchair. And when she finally returned to her family, it was not in her home but in a refugee camp, where the conditions were deplorable.

Mina had weathered all this adversity, but witnessing the stabbing seemed to be the last weight she could psychologically bear. After the violent attack, she became extremely agitated and wouldn't calm down. She was screaming, shaking visibly, and repeatedly stating that she didn't want to die.

Mina eventually regained her composure, but her behavior didn't return to normal. Instead, in the days following the stabbing, she became very withdrawn to the point where she stopped talking. Then, she shut her eyes and ceased responding to the outside world. She lay in bed in what appeared to be a semi-comatose state. She swallowed when her father hand-fed her, but that was the extent of her interaction with the environment.

After about a month, Mina was sent to a hospital in Athens; upon admission to the hospital, she was still mute and nonresponsive. Her vital signs, however, were normal, as were her reflexes. Doctors tried to elicit pain to test the extent of her impaired consciousness by doing things such as pressing down hard on the bed of her fingernails or pressing their fingers inward at the side of her jaw (both common methods of testing pain reactions in unresponsive patients). When these things are done to patients in a coma, they generally do not respond at all. Mina, however, displayed a painful reaction, suggesting that she wasn't in a coma—at least not in the typical medical definition of the word.

Nevertheless, several more months went by, and Mina remained in her sleep-like state. By the beginning of February 2020, she had been nonresponsive for over five months. Then, after an uncomplicated surgery to resolve what seemed to be an unrelated issue, Mina started to reawaken to the world. Within a week after the surgery, she opened her eyes once again. She began to develop an awareness of her environment and soon she was having conversations with her family. Mina claimed to have no memory of the six-month ordeal she had just been through.[28]

Mina experienced what doctors call a *functional coma*. She did not undergo the type of trauma, disease, or brain damage that typically leads to a coma, and doctors couldn't find any physiological cause for her state of unresponsiveness—but by all indications it was completely involuntary.* The surgical intervention inadvertently seemed to help bring Mina out of her comatose state, yet doctors are unsure why it did.

Mina's functional coma was a manifestation of a puzzling condition known today as a *functional neurological disorder*, or FND. FND didn't become an official diagnosis until 2013, but it's a new name for a condition that has gone by several different names over the years, such as *hysteria, conversion disorder,* and *psychosomatic illness*. A patient with FND experiences symptoms—such as weakness or paralysis, tremors, visual disturbances, and even seizures—that are typically caused by neurological dysfunction. But in an FND patient, these symptoms aren't clearly linked to any identifiable disease process affecting the nervous system.

Research suggests that the symptoms FND patients experience are very real, but because the origin of their symptoms has been so difficult to elucidate, for years patients with what is now known as FND were dismissed as having symptoms that were "all in their heads." Countless FND patients likely experienced this type of insensitive treatment, as studies suggest more than one-third of neurology patients have symptoms that cannot be traced back to a known disease,[29] and FND is among the most common diagnoses made at neurology clinics today (often second only to headache).[30] One shudders to think of the accumulated psychological stress caused

---

* Interestingly, physicians have noted that this type of withdrawn state is more likely to occur in traumatized children and adolescents, and especially in refugees like Mina. The poorly understood condition has been termed *resignation syndrome*.

by doctors who were indifferent to a patient's symptoms simply because they didn't neatly line up with an existing diagnostic category. Nevertheless, recent research is beginning to identify some distinct aberrations in brain function in FND patients, which is helping to convince doctors to treat the condition more like any other neurological disorder.

FND patients, for instance, display abnormalities in their sense of *self-agency*, a term that describes the perception that we are the ones creating our own actions or thoughts. This may cause FND patients to perceive certain bodily actions as involuntary, even if they aren't truly outside the realm of their control.

For example, some FND patients can develop a tremor that doctors can't find a cause for, known as a *functional tremor*. Studies, however, have found that brain activity during such functional tremors occurs in pathways involved in voluntary movement,[31] suggesting that patients are initiating these tremors—even though they are unaware they're doing so. This type of impairment in the sense of self-agency could also underlie other aspects of FND, such as the persistent inhibition of movement seen in Mina's case.

Additionally, many patients with FND have deficits in regulating emotions. This may involve a tendency toward strong emotional responses, along with difficulty suppressing unwanted emotions. In the brain, these types of dysregulated emotional reactions are linked to increased activity in areas such as the amygdala, and reduced activity in the prefrontal cortex, which (as we saw in the previous chapter) can exert inhibitory control over the amygdala to rein in excessive emotion.[32]

It's thought that the strong emotional reactions in FND patients may interfere with the brain's ability to make accurate determinations about the capabilities of the body. The brain thus erroneously assumes that some aspect of bodily functioning is impaired, and that assumption is so powerful it can override signals coming from the body that suggest everything is working as it should. The result is that the brain perceives the body as dysfunctional, despite there being no real physical impairment. And the brain prompts the body to act in accordance with its presumptions.

In Mina's case, for example, extreme psychological stress linked with a disrupted ability to regulate emotions may have caused her brain to make a flawed assumption about how her body should work. Then, her brain coerced her body into corresponding with its predictions. Perhaps the extremity of Mina's response aligned

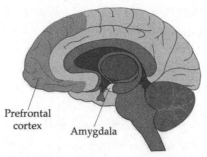

Prefrontal cortex

Amygdala

Note that this view of the prefrontal cortex is different from previous views of the prefrontal cortex in the book because this image shows the middle of the brain, as if the organ were split in half (this is necessary to see the amygdala, which is not visible from the brain's surface). Earlier images of the prefrontal cortex showed the region in a whole brain.

with the severity of the trauma she experienced, causing her body to shut down on a long-term basis.

There is a lot of speculation involved in hypotheses about what's going on in the brain in FND, but the condition is representative of a push in neuroscience—and health sciences in general—to appreciate the inseparable and mutually influential relationship between the brain and body. Gone are the days when we would talk about the health of either in isolation. A robust connection between the two is integral to our ability to make sense of not only what happens in the world around us, but also what happens within the confines of our own skin.

Cases like Mina's also underscore how important beliefs can be to the way our brains and bodies work, as FND represents what could be considered an extreme version of disrupted belief. In

FND, patients lose the ability to develop accurate beliefs about things such as their self-agency and the nature of their physiological functioning, and this has a serious impact on health. Thus, as we've seen in this chapter, what you believe can be tied to a surprising array of outcomes, such as improved results after taking a drug, an impaired sense of how much control you have over your body—and maybe even how long you live.

8

COMMUNICATION

On a sunny April morning in the early 2010s, Arnav sat down to read the newspaper with a hot cup of tea—a daily ritual he had practiced for well over a decade. He cracked the window next to him to let in a little air; it seemed like today would be a hot day in Delhi, just like the day before, and the day before that. Nothing about the morning so far suggested it would be particularly unique.

When Arnav unfolded his newspaper, however, things suddenly felt jarringly surreal. He blinked at the words on the page as confusion—and then panic—overtook his thoughts. *He could not read.* Arnav was 55 years old and a college graduate. He had been reading for 50 years, but this morning the words on the page made no sense to him.

He set the newspaper down, walked over to the sink, and splashed some cold water on his face. He rubbed his eyes and paced around the kitchen. Maybe this was transient—something that would fade as quickly as it appeared, one of those aberrations of brain function that became more common the older he got.

But when he sat back down and picked up the newspaper again, he still could not read the words in front of him. He tried to focus on letters. He found that he could read individual letters. But he could not string them together to make words.

Arnav got up and went outside, thinking that going for a walk might be a good way to clear his head. But as he walked, he saw street signs, shop signs, and words on clothing worn by passersby, and he became even more unnerved to realize his new impairment was not confined to newsprint.

Dismayed, Arnav went to the hospital. Doctors initially thought he must have some sort of visual disturbance. But the ophthalmologist declared Arnav's vision to be normal and sent Arnav to neurology.

When a neurologist examined Arnav, she found a new twist to his condition. To test the extent of his deficit, the neurologist gave Arnav a pencil and asked him to write out an explanation of what he was experiencing. Arnav initially thought the request was silly—he couldn't read, how would he be able to write? But when he picked up the pencil, the writing came surprisingly easy to him. He quickly wrote down, "I am able to write but unable to read."

When the neurologist asked Arnav to read what he had just written, however, he could not. Realizing Arnav was suffering from more than just a visual defect, the neurologist sent him for an MRI, which revealed that Arnav was dealing with the aftermath of a stroke.[1]

Writing without reading

The acquired inability to read (not due to a visual impairment) is called *alexia*, and it's a potential, albeit uncommon, consequence of a stroke. Typically, patients with alexia also have trouble writing— a condition known as *agraphia*. Given that Arnav's writing ability was preserved, his deficit is called *alexia without agraphia*.

Alexia without agraphia is rare, and there are still some questions about what causes it. But the leading hypotheses suggest it involves a disruption in visual input to a brain region called the *Visual Word Form Area*, or VWFA, which is thought to be critical to the recognition of words. The VWFA consists of a small section of cerebral cortex situated toward the back of the brain, near the junction between the occipital and temporal lobes. Since the early

2000s, researchers have accumulated evidence that suggests the VWFA is active during reading but not during writing or hearing spoken words.[2]

In order for the VWFA to do its job, it gets visual information from a part of the occipital lobe known as the *visual cortex*, which is the location of the main visual-processing areas of the brain. Thus, when we look at a word, the visual cortex generates an image of the word and sends the information to the VWFA, which helps to identify the word.

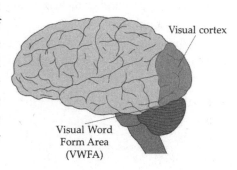

Visual cortex

Visual Word
Form Area
(VWFA)

Arnav's stroke had been caused by the blockage of an artery called the posterior cerebral artery, which is the main supplier of blood to the occipital lobe. Therefore, Arnav had lost the blood supply to his visual cortex, rendering the region partially nonfunctional. Consequently, Arnav's visual cortex could not provide the VWFA with visual information, and the area of his brain that would normally translate the words on the page was never "seeing" what those words were—causing him to be unable to read. Parts of his brain involved in language production, however, still maintained connections with areas needed to make the movements required for writing, which enabled him to retain his writing ability.

Changing perspectives on language function

The VWFA's hypothesized specialization for one aspect of language (i.e., reading) underscores an important point about the neuroscience of language: it is a complicated function made up of a catalog of individual tasks. Just speaking a simple sentence, for example, requires the successful execution of operations such as word retrieval, the application of syntax (i.e., the rules used to properly arrange words in a sentence), coordinating the activity of the muscles involved in speech, sprinkling in appropriate changes in tone and pitch, and so on. Each of these tasks might require the contribution of different parts of the brain, causing language to be reliant on a large number of functioning brain regions for it to be fully operational.

This, however, is not how neuroscientists have historically viewed language. Instead, attempts to understand how language is processed and produced by the brain have—until relatively recently—been dominated by a perspective that involves two main areas: one for language production and the other for language comprehension. These two areas, known as *Broca's area* (for language production) and *Wernicke's area* (for language comprehension), are named for the nineteenth-century neuroscientists who discovered them: Paul Broca and Carl Wernicke, respectively. Broca's area and Wernicke's area are connected by a bundle of neurons called the *arcuate fasciculus*,* and according to the traditional (but now outdated) neuroscientific perspective on language, the resultant system can explain most of the critical features of communication.

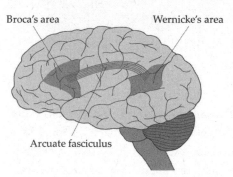

Broca's area   Wernicke's area

Arcuate fasciculus

Brain areas involved in the classic (but now outdated) model of language.

In this model, Wernicke's area extracts meaning from language that is heard and adds meaning to intended speech so it makes sense. Broca's area, on the other hand, is involved with

---

* The term *fasciculus* means "bundle" and is sometimes used in neuroscience to refer to a bundle of neurons. *Arcuate* means "curved," so the term roughly translates into "curved bundle."

the stimulation of the parts of the brain that can activate muscles required for speech (e.g., mouth, throat, respiratory muscles). The arcuate fasciculus connects the two areas and enables them to work together to produce and understand language.

Today, however, this model is considered woefully incomplete. Part of the reason for this assessment is that modern neuroscience has uncovered a multitude of additional brain regions (such as the VWFA) that are involved in language. In many cases, these other regions make very specific contributions to language.

At the same time, I don't want to give the impression that language is the result of a collection of brain areas all working independently. On the contrary, these disparate regions interact with one another continually, and that communication is essential to healthy language function. Thus, neuroscientists now see language as an undertaking that requires the cooperation of vast networks of neurons. An interesting consequence of this network approach to language is that damage to a single component of the network may result in the loss of one particular element of language, resulting in some unique deficits.

Herpes and the brain

Mika was 25 years old in 1995 when she was admitted to the hospital in late summer with a fever and severe sleepiness.[3] Sleepiness during a fever is, of course, not uncommon, but with Mika it was the degree of her somnolence that was concerning. She could barely stay awake to conduct a brief interview with a doctor, which suggested Mika was experiencing an affliction more serious than a simple flu.

Doctors suspected that something was compromising Mika's neural functioning, so they used MRI to obtain an image of her brain. They found that she had evidence of damage to her temporal lobe, primarily in her left cerebral hemisphere. Further tests indicated that Mika's brain damage was likely caused by a condition known as *herpes simplex encephalitis*.

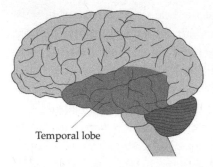

Temporal lobe

Herpes simplex is a viral infection caused by the herpes simplex virus; it can result in a range of symptoms—the most common of them being the well-known sores that develop around areas such as the mouth (cold sores) or genitalia (genital warts). The herpes virus is incredibly common; indeed, more than 60 percent of adults under age 50 carry the form of the virus that typically causes symptoms like cold sores, and more than 10 percent of that same age group carries the form of the virus that causes genital warts.[4] The presence of the virus, however, is not apparent in most people at any one point in time, as herpes has a remarkable ability to conceal itself within your body, waiting for an ideal opportunity to awaken and spread to another host. Herpes is a virus that really should make you think twice about who you kiss...or do other "things" with.

Typically, herpes hides in sensory neurons close to the site where infection originally occurred, but sometimes it also uses those neurons to maneuver its way to the brain. In rare cases, herpes infects the brain and causes brain inflammation, also known as *encephalitis*.*

Inflammation is a standard reaction of your immune system to any infection or injury. It's the reason why when you drop something on your toe, your toe becomes swollen, red, and tender. These symptoms are unintended consequences of the immune system rushing blood rich in immune cells to a location where tissue damage (and a higher potential for infection) is present, and then waging a war there against any invading germs.

---

* While a number of viruses can cause encephalitis, herpes simplex is the most common cause of encephalitis in the United States, accounting for thousands of cases of encephalitis each year in the US.

Pathogens in the brain can also prompt an inflammatory response, which has unintended consequences of its own. These include collateral damage to neurons—which end up being the victims of an overzealous immune system focused on eradicating the invading germ at all costs—and swelling of brain tissue (known as cerebral edema), which can further put neurons at risk. Add these effects of inflammation to the damage caused by the pathogen itself, and a simple infection with something like the herpes virus can be devastating—and even deadly.

In Mika's case, the herpes virus had migrated to her brain, and the resultant infection and inflammatory response had already killed a substantial number of neurons. Doctors gave Mika an intravenous antiviral medication, which fought off the infection long enough for her acute symptoms to diminish. After about a month of treatment, she was released from the hospital.

But Mika's battle with the virus didn't end there. Two years after her first hospital stay, she was back in the emergency room with a high fever, intense headache, and some concerning signs that the virus was still affecting her brain. Doctors did another MRI of Mika's brain and found the virus had wrought even more damage. Now her left temporal lobe was severely atrophied; the amount of neuronal loss was significant.

Additionally, Mika was displaying a new and unusual language deficit. She could speak fluently for the most part, understand all that was said to her, and read and write without any trouble. But she was severely impaired in recalling the names of things.

For example, when Mika was shown a picture of a pair of shoes and asked what they were, she could not remember the word *shoes*, and instead described them as "things people usually put on when walking." The same deficit occurred when doctors asked her to identify almost anything: animals, food, clothes, vehicles, etc. She could not remember what these things were called, although she could describe them in detail otherwise.

Mika was suffering from a condition known as *anomic aphasia*. *Anomia* literally means "without names," while *aphasia* is a more general term used to refer to a language disorder. Patients with anomic aphasia have a specific deficit in word retrieval that is typically most pronounced with attempts to recall nouns and verbs. Often, these individuals can describe objects and use hand gestures to aid in their descriptions, so they usually are still able to communicate—albeit in a disrupted manner.

Anomic aphasia can be caused by damage to various regions of the cerebral cortex—which part of the cortex specifically depends on the deficit the patient is experiencing. Patients who have trouble retrieving verbs are more likely to have damage closer to the frontal parts of their cerebral cortex, while those who experience difficulty retrieving nouns tend to have damage somewhere in the temporal lobe. Even finer distinctions can be made as well, where damage to a certain part of the temporal lobe increases the chance there will be impairments in the naming of objects, damage to another area is more likely to cause deficits in the naming of living things, and so on.[5]

The diversity of language dysfunction

Anomic aphasia and alexia without agraphia are just two of many distinct language disorders that can emerge due to some abnormality in brain function. Other patients know exactly what they want to say but can't get the muscles involved in speech production to work properly—an impairment known as *apraxia of speech*. They have difficulty pronouncing words they know, frequently make errors in speech, and struggle to formulate words. They often have to attempt to say a word multiple times before it comes out correctly. The deficit, however, is solely in the movements associated with speech (which are, by the way, more complicated than you might expect: there are over 100 muscles involved in speaking, including muscles that move your lips, tongue, throat, cheeks, and jaw).

Some patients have the opposite problem: a surplus of language. They might experience *paraphasia,* where speech is effortless but riddled with unintended syllables, words, and/or phrases; these superfluous language elements often make speech nonsensical. Or they might display a verbal tic known as *echolalia,* in which they unintentionally repeat words—either others' or their own. And even if you don't know it by name, you're likely familiar with *coprolalia,* which involves the involuntary uttering of obscene or inappropriate comments, because you've probably seen someone with this symptom portrayed in a movie or television show—or perhaps you're familiar with one of the many people who document their struggles with coprolalia on TikTok or other social media. Although many people associate coprolalia with Tourette's syndrome, coprolalia appears in fewer than 20 percent of cases of Tourette's.[6] It can, however, also occur in other conditions and is a rare consequence of stroke.

People may experience a similar range of problems with written communication. Damage to language areas of the brain can cause someone to suddenly be unable to communicate through writing or make them incapable of understanding written language. Or, analogous to paraphasia, they may unintentionally insert unnecessary words, syllables, or letters into writing—even while verbal communication remains normal.

For example, a condition known as *hypergraphia* involves a tendency toward rambling and excessive written communication. One stroke patient with hypergraphia who was asked by his doctor how he was feeling picked up a pencil and paper and began writing, "Do not suggest. You are unfair. I don't care if you know you are fair or not. You should use your suggestions for a better purpose." He continued in this fashion until he had filled three more sheets of paper with such digressions. On another day, doctors asked him to write only his address on a piece of paper. He ignored the request and wrote a three-page description of how his illness had occurred.[7] Often, these patients display normal speech otherwise; they can have a typical conversation when talking but ramble inordinately when writing.

A list of all the potential disruptions to language that can occur after some neurological insult is truly extensive, underscoring the complexity of language and the brain networks that contribute to it. Neuroscientists now recognize that these language networks span the entire brain, linking regions throughout the organ. But, as I've mentioned, the history of the neuroscience of language has been rife with misunderstandings, and until fairly recently it was thought that one half of the brain made little—if any— contribution to language at all.

The divided linguistic brain

Since early studies on the neuroscience of language, one curious observation has been made repeatedly: in most people, language is especially dependent on activity in the left hemisphere of the brain. Paul Broca (the nineteenth-century neuroscientist mentioned earlier in the chapter) first recognized the predominance of the left cerebral hemisphere's role in language in the mid-1800s when he found a reliable pattern in which patients who had lost the capacity for speech also had damage to the left sides of their brains.[8] This was a surprise to Broca, as up to that point it was widely assumed that both halves of the brain were identical in structure and function.

But Broca's research suggested the left cerebral hemisphere played the more important—or "dominant"—role when it came to speech. Subsequent evidence has continued to support this perspective into the present day. Damage to the left cerebral hemisphere (e.g., from a stroke) carries a high risk of causing some substantial disruption to either the formulation or comprehension of language. Damage to the right cerebral hemisphere, although it's linked to problems of its own, is less likely to cause a language disorder.

Due to these observations, for years the right hemisphere was described as nonverbal. But as our ability to probe the functions of the brain has become more refined, researchers have realized that the right cerebral hemisphere has a larger role in language than initially suspected. We now know, for example, that the right hemisphere does have the capacity to understand language. Moreover, it's critical for some of the more subtle aspects of language, such as the incorporation and understanding of tone and rhythm in speech, a characteristic known to linguists as *prosody*.

Without prosody, our speech becomes devoid of any inflection or change in emphasis. One consequence of this is that language loses its ability to convey emotion. Take, for example, the case of Charlie, a 63-year-old man who suffered a stroke that caused substantial damage to his right cerebral hemisphere.[9] After the stroke, doctors immediately noticed something wasn't quite right about Charlie's linguistic tendencies. He talked in monotone, detached and emotionally unaffected by what he said—more like an objective narrator of his life than a participant in it. He also didn't use any hand gestures.

Charlie's emotional detachment became even more apparent when he talked about things that would arouse feeling in most people. For example, Charlie recounted his military experiences liberating concentration camps in Germany as flatly as if he were describing a dental cleaning. He discussed the murder of his son, who was shot to death the previous year, as if it were no more significant than a trip to the grocery store.

Charlie couldn't even mimic emotions such as sadness or anger when he tried. Doctors asked him to, and Charlie just spoke louder—but still in a flat and impersonal way.

Charlie's condition is known as *aprosodia*, and patients with it can generate speech but cannot add emotional inflection to it; most can't detect or understand emotion in other people's voices, either—even though their ability to experience emotions may remain intact. This severely impairs their ability to participate in meaningful conversations, as emotional tone plays an incredibly large role in interpersonal interactions.

Moreover, patients with aprosodia have trouble with all aspects of prosody, not just emotion. They're impaired when it comes to producing language whose meaning is dependent on timing, pitch, or volume, and they struggle to comprehend these elements in others' attempts to communicate with them as well. This leads to a general deficit in communication, which underscores the importance of prosody in language.

Think about the times you have implied a question just by raising the pitch of your voice at the end of a sentence (something linguists refer to as the *high rising terminal*). For example, imagine you ask, "There's no coffee being served at this meeting?" with your voice rising at the end to clearly indicate a question. To someone with aprosodia, the change in pitch isn't detected, so neither is the question—it just sounds like a statement. They would think you were telling them there's no coffee being served, not asking them about it. Similarly, differences in emphasis that indicate word choice ("obJECT, as in "I obJECT" vs. "OBject," as in "what is that OBject?") are indistinguishable, as are differences in emphasis for other reasons (e.g., "I ordered COFFEE" when the waiter brings you tea). Sarcasm, of course, is a lost cause.

Aprosodia is typically the result of damage to the right cerebral hemisphere, which (along with other evidence) implicates the right half of our brain in prosody. Thus, language researchers now acknowledge that the right hemisphere plays important roles in various aspects of the nonverbal expression of language (such as changes in pitch and tone). But more recent research also points to contributions from the right hemisphere in "typical" language tasks such as comprehension, language acquisition, word recognition, and so on.[10]

The right cerebral hemisphere is therefore now appreciated as playing a significant role in language—even if its contributions are different from those of the left hemisphere. Nevertheless, some of the most recognizable language impairments are still seen after damage to the left side of the brain. And, while the right hemisphere is especially important to prosody, damage to the left hemisphere can also disrupt the patterns of speech production— sometimes in an even more unusual fashion.

Waking up with a foreign accent

In 2009, Karen Butler had dental implant surgery, a common procedure to permanently replace missing teeth lost for any variety of reasons ranging from old age to injury. Rather than be put completely under for the procedure, Karen elected to have "twilight anesthesia," which involves the use of medication to put the patient into a sedated state where they are conscious but have little recollection of the experience.

The surgery seemed to go well; Karen's dentist didn't note any difficulties with it. When the anesthesia started to wear off, Karen's mouth was sore and swollen, which was to be expected. Karen also noticed, however, that her voice sounded…funny. Her dentist assured her that this was due simply to the swelling, and that her voice would return to normal in a few days.

But it didn't. After Karen's swelling went down, her voice still sounded very strange. Karen had been born in Illinois and raised in Oregon (where she lived at the time of the surgery). She had left the United States only a handful of times to make trips to Mexico and once to Canada. But after the surgery she was unmistakably speaking with an accent.

Karen expected her newfound—and completely involuntary— accent might fade with time. But it never has, and she has simply been forced to get used to it. Her accent isn't a clear reproduction of that of native speakers of any one country. In fact, it sounds a bit like a mixture of British, Irish, and maybe even some Transylvanian. But one thing's for sure: compared to the accents of her fellow Oregonians, Karen sounds like a foreigner.

As shocking as it was for Karen to develop a new foreign accent, she has learned to like it. She says it can be a great conversation starter and has helped to transform her from a shy, introverted person to someone who is more outgoing and comfortable talking about herself.[11]

Karen has been diagnosed with a rare condition known as *foreign accent syndrome,* or FAS. FAS was first recognized in the early 1900s, and since then doctors have only seen just over 100 cases of it.[12] Typically, FAS patients develop what sounds like a foreign accent after suffering some sort of brain trauma, such as a stroke or head injury. In Karen's case, doctors think it's possible she had a small stroke while under the influence of anesthesia—but a stroke mild enough to leave her with no lingering symptoms except her newly acquired accent.

In most cases, patients who develop FAS have not had previous exposure to the accent they adopt.[13] In fact, patients with FAS aren't truly speaking with a foreign accent at all, as the "accent" does not consistently mimic that of an established language. Instead, it seems to be the result of a combination of impairments that interfere with articulation, timing, prosody, and other speech patterns in such a way as to give the *impression* that the speaker has developed a foreign accent. FAS is frequently associated with damage to regions of the left cerebral hemisphere that are involved in speech-related movements, such as the activation of the muscles in the larynx (i.e., voice box), and movement of the tongue and lips.[14]

In some patients with FAS, however, their presentation of symptoms doesn't mesh with the idea that their condition is caused purely by brain damage. For example, their accent can be inconsistent—applied to certain words but not to others or undergoing abrupt changes that drastically alter its characteristics. And in some cases, patients display abnormalities in speech that linguists recognize as signs someone is intentionally speaking with a foreign accent, such as occasionally (but not consistently) leaving the -ing ending off of verbs or adding an "s" sound to some words for no apparent reason.[15]

For these cases, researchers suggest there is something more than just a neurological deficit going on. Instead, psychological factors, often linked to a psychiatric condition—or at least some exacerbating situation such as severe stress—are thought to be prompting the changes in speech. It doesn't mean these patients are faking their accent, but instead that they are experiencing some psychological impairment that can't be clearly traced back to brain damage. Indeed, in some of these cases, there may have been brain damage preceding the initial development of the accent, but eventually psychological mechanisms could have taken over and propagated the speech abnormalities.

Whatever the case, FAS exemplifies just how strange disturbances to typical language function can be. As illustrated by the cases in this chapter, damage to the brain can cause almost any aspect of language to be lost or altered in ways that can be bewildering for the patient and those around them. Thus, while our rich and expressive language is one of the most impressive human accomplishments, its dependence on the brain also makes it incredibly vulnerable.

9

SUGGESTIBILITY

Linda didn't think she could take living in the apartment she shared with her mother anymore. For three years, her next-door neighbors on both sides had been tormenting her and her mom by playing music and sound effects—at high volume— 24 hours a day. The neighbors on her left, Linda claimed, played songs such as the Irish ballad "Danny Boy" throughout the day and night. Those on her right played a recording of a baby crying nonstop.

Linda and her mother made repeated efforts to put an end to the harassment, which frequently woke them up in the middle of the night and filled their days with clamor and distress. At first, they simply complained and asked the neighbors to stop. But those requests were ignored, so Linda and her mom began pounding on the bedroom walls when the music or sounds got too loud (which was often).

Linda was desperate and asked her sister Jodi to intervene on her behalf. Jodi talked to Linda's neighbors, but rather than ending the dispute, Jodi came back convinced the neighbors weren't doing anything wrong. In fact, she said, the neighbors were flabbergasted—they claimed they weren't playing any music nor sound effects and argued that Linda and her mother were the unruly ones for pounding on the walls at all hours of the night. Worst of all, Jodi seemed to believe the neighbors over her own sister.

After the attempted intervention from Linda's sister, the neighbors only got louder. It was as if they intended to punish Linda for involving someone else in their dispute. Now, Linda could hear the music even when she wasn't at home—sometimes when she was miles from the apartment.

But of course it would have been impossible for Linda to hear music at such a distance, so you might be able to see where this is going. Linda eventually ended up in the hospital, where doctors determined she was experiencing delusions and auditory hallucinations. The sounds that Linda was tortured by weren't being made by her neighbors at all; they existed only in her head. But that didn't explain why her mother heard the same things.

In fact, Linda's mother had been the first one to hear the sounds coming from their neighbors' apartments, and she had started waking Linda up in the middle of the night to alert Linda to the disruptive noises. Initially, Linda didn't hear anything. But after her mother's repeated exhortations, Linda began to hear the sounds so distinctly that she could sing along with the music.

Strangely enough, during a six-month period when Linda's mother was out of town visiting relatives in Scotland, Linda stopped hearing the sounds—but Linda's mother did not. She still heard them—and still blamed the neighbors back home for them—even though she was hundreds of miles away. Linda began hearing the sounds again just before her mother was due to come back from her trip.

Doctors eventually determined that Linda and her mother were suffering from a rare phenomenon known as *folie à deux*.[1] *Folie à deux*, which literally means "madness for two," involves cases where two people adopt the same delusional beliefs. The term was coined in the 1870s, but today it also goes by other names such as *shared psychotic disorder* or *induced delusional disorder*.

While *folie à deux* typically affects a pair of individuals, there have been cases where more people are involved, and sometimes whole families fall under the same delusion. When this happens, it's referred to as *folie à famille* ("family madness"). Frequently, in cases of *folie à famille*, the delusions become multigenerational (involving, for example, grandparents, parents, and children) as they are passed on like a family heirloom.[2] In any case, *folie à famille* can initially be baffling for medical professionals who are attempting to get to the root of a problem.

For example, the members of one family (we'll call them the Millers) all developed a condition known as *delusional infestation*, which involves the false belief that the body has become infested with some living or nonliving thing. Individuals with delusional infestation may come to think they are crawling with parasites, insects, worms, or other small creatures, or they might believe they are infested with a nonliving material, such as miniscule fibers, threads, etc.

The belief can become intense and all-consuming, and patients desperately seek out ways to rid themselves of the imagined scourge. They often use tweezers or needles to try to pick the violating substances out of their skin and might turn to insecticides, bleach, or other cleansers to eradicate the infestation.

The Millers' delusion began when Mrs. Miller developed a fear that neighbors and relatives were out to harm her and her family. Somehow (the details here are unclear even to Mrs. Miller), she believed these outsiders caused an infestation of parasitic worms in her household.

The incident came to the attention of doctors when Mr. Miller went to a dermatologist complaining of incessant itchiness all over his arms, abdomen, and back. The doctor noted that Mr. Miller had dry, red patches across his body that clearly had been irritated (or caused) by scratching. The dermatologist recommended several treatments, but nothing worked—and the doctor couldn't identify a cause of the itchiness. In the course of treatment, Mr. Miller mentioned that his wife and two daughters were suffering from the same type of irritation.

Mr. Miller went to several doctors, and none could find evidence of any sort of infestation. A psychiatrist finally made the only official diagnosis: the whole family was suffering from a shared delusion. Apparently, the delusion had begun with Mrs. Miller; her husband had initially resisted the idea. But Mrs. Miller was the dominant personality in the house, and eventually she persuaded Mr. Miller to share her delusional thinking. Soon after, both daughters bought into it as well. When doctors recognized the Millers as a case of *folie à famille*, they recommended psychiatric treatment—which the Millers adamantly refused.[3]

The recipe for a shared delusion

Although it's not completely clear how shared delusions take hold, it seems that suggestibility and personality dynamics play a prominent role. Typically, a shared delusion begins with the development of delusional thinking in someone who has great influence over another individual. The person who first experiences the delusion may, for example, be more intelligent or have seniority, which gives their opinions increased weight in the eyes of the person who eventually comes to share their delusion. Often, the pair or group lives in relative social isolation, which makes it less likely that the individual who develops an induced delusion will be enlightened by someone else offering a more rational perspective.

The initial delusions in a shared delusional disorder can arise due to conditions that are known for causing distortions in normal thinking, such as schizophrenia or dementia. But the person who develops the delusions secondarily is usually not suffering from an identifiable brain ailment. They may, however, exhibit a high degree of suggestibility as a personality trait. This suggestibility, combined with the undue influence of the primary patient—and a sprinkle of social isolation—could be the special ingredients that lead to the development of a shared psychotic disorder.

One hypothesis for what might go wrong in the brain to allow a shared delusion to take hold involves aberrant activity in brain regions responsible for generating doubt. You might remember from previous chapters that the right cerebral hemisphere is thought to play a role in checking the rationality of explanations for the events in our lives. Neuroscientists hypothesize that these plausibility-checking circuits may overlap with circuits that produce doubt when we encounter information that seems fishy, even if it's not blatantly impossible. Abnormal functioning in these doubt-generating circuits, then, might cause someone to be particularly gullible, and more likely to be unduly influenced by a dominant personality.

Some research suggests this doubt-generating circuitry is found in the prefrontal cortex, as studies indicate that the region plays an important role in suggestibility and credulity. According to this

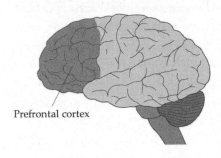

Prefrontal cortex

research, when we encounter information that's questionable, the prefrontal cortex is involved with "tagging" that information as potentially false. But the proficiency of the prefrontal cortex in this task varies from person to person. A child's prefrontal cortex, for example, is not yet fully developed,* and elderly individuals often have experienced some degree of age-related atrophy of the prefrontal cortex. And, of course, there are other factors (such as genetics, early childhood influences, and drug and alcohol use) that can affect prefrontal cortex functioning.

These variations in how the prefrontal cortex works might lead to differences in a person's ability to generate doubt about improbable scenarios. In one study, for example, scientists worked with patients who had experienced some sort of damage to the prefrontal cortex (e.g., due to a stroke). To gauge the participants'

---

* Surprisingly, the development of the prefrontal cortex may not be complete until someone reaches their mid-20s. Because the prefrontal cortex is involved in decision-making and impulse control, its delayed maturation may help to account for the impulsive and sometimes reckless behavior exhibited by adolescents and young adults.

ability to evaluate dubious information, researchers showed them magazine or newspaper advertisements that had been doctored to be obviously misleading. For instance, one of the advertisements presented a pain reliever that claimed to be able to get rid of headaches "without the side effects of over-the-counter pain relievers." However, a noticeable disclaimer at the end of the ad stated, "This product can cause nausea in some consumers when taken regularly."

The participants with prefrontal cortex damage were significantly less likely to be able to identify the misleading advertisements compared both to participants with damage to other areas of the brain and participants without any brain damage. Damage to the prefrontal cortex was not, however, linked to poorer cognitive performance (compared to other brain-damaged patients), so an overall deficit in cognitive function couldn't explain the differences. The researchers thus hypothesized that specific regions of the prefrontal cortex play a critical role in identifying information as potentially false or misleading.[4]

Hypnosis, suggestibility, and the prefrontal cortex

More support for the idea that prefrontal cortex functioning is important to suggestibility comes from research into the neural basis for hypnosis, a practice that by definition involves a state of heightened suggestibility. For many of us, the word *hypnosis* conjures up images of a monocled Freudian therapist swinging a gold watch back and forth before a patient's eyes and repeating, "You're getting sleepy," in a heavily accented voice. True hypnosis is a more subtle experience that uses relaxation techniques and visualization to help a patient adopt a mindset that is receptive to change.

In many patients, hypnosis does lead to a state of increased suggestibility; someone trained in the therapeutic applications of hypnosis can utilize that suggestible state to alleviate symptoms of depression or anxiety, mitigate pain, aid in smoking cessation, and so on.[5] Hypnosis has even successfully been used in place of anesthesia for some types of surgery[6]—a select few hospitals are beginning to incorporate it for certain procedures as a safer alternative to traditional anesthesia.[7]

Hypotheses about how hypnosis affects the brain often point to a role for the prefrontal cortex. Research, for example, generally supports the idea that hypnosis involves decreased prefrontal cortex activity, which makes someone more open to hypnotic suggestions.[8]

Researchers in one study tested this hypothesis using transcranial magnetic stimulation, or TMS. Recall that this is the procedure we talked about in Chapter 4 that involves noninvasively altering brain function by exposing the brain to magnetic waves. These magnetic waves induce electrical currents in the brain that can transiently disrupt the functioning of neurons.

To investigate the role of the prefrontal cortex in hypnosis, researchers applied TMS to a specific part of the prefrontal cortex in a couple dozen participants. Immediately afterward, they hypnotized the subjects and recorded how responsive they were to suggestions in a hypnotic state. The participants who had received TMS to the prefrontal cortex (decreasing activity there) were more likely to respond to hypnosis. More of them accepted hypnotic suggestions such as, "your arm is so rigid that it cannot be bent," or "you are experiencing a sour taste in your mouth."[9] These results imply that reduced activity in the prefrontal cortex might be linked to hypnotizability and suggestibility in general.

Thus, perhaps disrupted prefrontal cortex function also plays a role in the type of credulity that can lead to a shared delusion. At the same time, shared delusions involve more than just suggestibility; they also are heavily dependent on personality dynamics, typically occurring when someone in a dominant position exerts influence over another who is easily swayed. To gain more insight into this type of relationship, we can look to other examples of the power of social influences on the brain and behavior.

An undue influence

In 1955, a charismatic young preacher named Jim Jones established a church in Indiana that would eventually become known as the Peoples Temple (the omission of the apostrophe here is, by the way, intentional—to refer to the peoples of the world instead of suggesting the people's ownership of the temple). After having a vision that Indiana was destined to be destroyed in a nuclear explosion, Jones relocated the church to a small town in Northern California, where he became a prominent and respected figure in the community. His message of social and racial equality attracted idealistic young people, and Jones's generous charity work made him well-liked. His church offered programs to aid those in need, including a soup kitchen, drug rehab, and free legal services.

The Peoples Temple grew along with Jones's reputation; he had around 20,000 followers by the early 1970s. But as the size of his congregation increased, so did the frequency of reports that the Peoples Temple had some unorthodox—and even downright abusive—practices. Disillusioned and disgruntled ex-members of the church told of services that lasted until dawn and public humiliations, including "spankings." One 16-year-old girl, for example, was paddled on the bottom with a large board 75 times in front of a congregation of around 700 people while her parents watched. She claimed she "couldn't sit down for at least a week and a half."[10]

These reports spawned investigations into the Peoples Temple, and those investigations suggested Jones was a demagogue who manipulated his congregation with intimidation and manufactured showmanship, such as fake healings. Facing increased public scrutiny, Jones took more than 1,000 of his followers and fled to the small country of Guyana in South America. There, Jones established a compound in the middle of the jungle; it would informally be called Jonestown.

In Jonestown, Jones ruled over his congregation like a despot. He had delusions of grandeur and even had a throne set up in the compound's main building. But along with his deluded belief in his own greatness came extreme paranoia and a deranged focus on isolating his community from any external influences. His mental health was declining rapidly, and his heavy abuse of amphetamines and barbiturates likely accelerated the descent.

Jones believed that interference from the outside world would lead to the dissolution of his community, and this ended up being a self-fulfilling prophecy. In late 1978, a California congressman named Leo Ryan came to Jonestown after complaints that family members of his constituents were being held there against their will. Ryan arrived with a group that included reporters and photographers. Uncertain what to expect, they were surprised when Jones welcomed them graciously. Ryan and his group even spent an evening having dinner and enjoying live music with Jones and his followers.

But the unblemished façade Jones attempted to construct for Ryan's benefit could not cover up the discontent and fear that had become widespread among the residents of Jonestown. Throughout the night, Ryan was repeatedly approached by members of Jones's congregation, who surreptitiously asked for help escaping from the oppressive community. Jones learned of these requests, which he considered treasonous, and decided he needed to act to prevent Ryan from leaving with any of the inhabitants of Jonestown.

Ryan planned to depart Jonestown the next day, and he agreed to take several of Jones's followers with him. As they were at the airstrip readying to board the plane, however, Jones's armed guards fired at them. The attack killed Ryan and four others.

Jones then made a decision that would turn Jonestown into one of the more gruesome events in US history. Jones believed that the United States government was due to arrive in Jonestown at any moment to put a definite end to his church and his community. He conveyed this to the residents of Jonestown, painting a picture of a governmental raid that would involve death and torture as punishment for attempting to live outside the boundaries of typical US society. The only way for his congregation to avoid this outcome, Jones proclaimed, would be to die on their own terms.

To accomplish this, Jones encouraged his followers to drink cyanide-laced Flavor Aid. Adults used syringes to squirt the poisonous concoction into their children's throats. Then, the adults drank the mixture themselves. In the end, over 900 were dead, including over 300 children. Jones died from a gunshot wound to the head, which was likely self-inflicted.

Cultish brains

Clearly, the degree of suggestibility exhibited by Jones's congregation is on a whole other level. What happens in the brains of such a large number of individuals to cause them to act in a way that seems guided by delusion? Do they all have some degree of disrupted prefrontal cortex functioning?

Maybe. One study found that people with damage to their prefrontal cortex were more likely to submit to authority, adopt dogmatic beliefs, and aggressively defend those beliefs.[11] Another study concluded that individuals with prefrontal cortex damage were less likely to develop a negative opinion of someone after learning they had engaged in immoral behavior.[12] Combining these types of deficits seems like the right mixture for someone who blindly follows the dictates of a cult such as the Peoples Temple.

But truthfully, we don't know what was happening in the brains of Jones's followers when they decided to give up their lives in the United States and move to Guyana, because scientists weren't able to study them. Thus, it's presumptive (and likely too simplistic) to attribute their behavior to one cause, such as prefrontal cortex dysfunction.

Cults* like the Peoples Temple may be more appealing to highly

---

* It's important to note that many researchers today shun the use of the word *cult*, as it has a pejorative connotation. As such, it may be misapplied to new religious movements that are not necessarily damaging or destructive to their followers. Regardless, I'll use *cult* here since I'm primarily referring to established destructive cults, such as the Peoples Temple.

suggestible people, but they also attract people who are emotionally vulnerable, have limited social support, have experienced some physical or sexual abuse, and/or are in otherwise desperate situations where resources—financial or emotional—are limited.[13] These individuals may be intrigued by joining a new community and being able to escape the troubles of their old lives.

At the same time, not everyone who joins a destructive cult like Jonestown is highly suggestible, in emotional distress, or coming from a desperate situation. Additionally, cult members tend to be better educated than the general public and disproportionately from middle- and upper-class households.[14] So, cults are not always full of the type of people you would expect to find in a cult; instead, they're full of people who believed—like you probably do as you're reading this—that becoming part of a destructive cult was something that could never happen to them.

Most people who join cults get inculcated slowly and insidiously. Their exposure to alarming practices is typically gradual enough that they can rationalize their continued involvement. By the time they realize what's happening, they're enmeshed in a new society that's difficult to break away from. They find themselves surrounded by others who have already gone through the indoctrination process and who may have become ardent advocates for the cult simply to justify their allegiance to it. The social pressure to remain part of the cult becomes a dominant factor in explaining why people have a difficult time leaving once they've begun to feel like part of the community.

Thus, cults provide an example of just how strong social influences on human behavior can be, and they hint at a feature of the human brain that affects all of us every day: it's more inclined to follow than lead. It relies on information from other people to help it to determine the correct thing to do, and sometimes this approach breeds behavior that is mistake-prone, if not disastrously misguided.

How peer pressure changes our minds

The Polish-American psychologist Solomon Asch demonstrated this dependence on social information in making decisions—and how it can lead to egregious errors—with one of the best-known experiments in the history of psychology. Asch was a social psychologist, and as such he was interested in how social factors, such as interpersonal and group dynamics, influence behavior. In the 1950s, he began to focus specifically on how peer pressure and the desire to conform affect the way we act.

In Asch's most notable experiment, he showed groups of college students (about eight at a time) a card with a line of a specific length—ranging from 2 to 10 inches—printed on it.[15] He then presented the students with another card with three

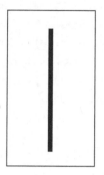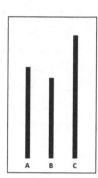

A replica of the cards used in Solomon Asch's experiment. Participants had to identify which of the three lines in the card on the right matched the length of the line in the card on the left.

lines of varying lengths on it, and he asked which of the three lines on the second card matched the length of the line on the first card. The answer was obvious: one line was identical, while the other two were off by anywhere from one-half inch to two full inches.

The students had to state their answer aloud, in front of the group. They gave their responses in order, one at a time, and the second-to-last participant to voice his decision just so happened to be the only person in the group who was not working with the experimenters (this was, of course, unknown to him*).

For the first two pairs of cards shown, everything seemed

---

* One of the main criticisms of this experiment has been that all the participants were male. Since Asch's original experiment, however, other comparable normal enough. All the students agreed on the lines that matched in length (you would almost need to be visually impaired not to). But to the shock of the one "true" participant, on the third card—and 11 of the 15 cards after that—everyone who answered before him made the same incorrect choice. Thus, the student was faced with a decision: give the answer that was obviously the right one—and in doing so contradict everyone else who had answered before him—or go along with the crowd even though it would require giving a blatantly wrong answer.

Most of the time, the true participants stuck to their guns and gave the answer they felt was correct. But 75 percent of them succumbed to peer pressure and chose an incorrect answer at least once, and about one-third of all the responses given by the true participants were wrong—which is quite high considering how obvious the correct answers were. When the true participants made an incorrect choice, it always erred in the direction of the answers given by the majority.

The experiment also had a control condition in which the participants wrote down their answers (so no one knew what they had chosen). Errors occurred in this situation less than 1 percent of the time because the true participants didn't feel the social pressure associated with openly stating something that contradicted the opinions of everyone else in the group.

Asch's experiments show that our brains highly—even inordinately—value social information. According to scientists, there is a good reason for this. For most of our history as a species (indeed, over 90 percent of it), we lived as hunter-gatherers. We wandered across the savannah in groups of a few dozen or so, and our survival depended on our ability to cooperate with those in our group.

Thus, it makes sense that our brains evolved to be attuned to social cues, adept at communication, and inclined to interact

studies have included female participants and observed that females display a similar, if not greater, tendency toward conformity.

with others. At the same time, our brains became wired to place a great deal of importance on socially obtained information, and to especially value ideas held by a large number of people.

Again, this is a rational strategy. Our brains developed a short-cut to evaluating information: the more people who believe something, the more likely it is to be true. While we are all aware of the problems inherent with this approach—think 1940s Germany for an obvious, if overused, example—the strategy points us in the right direction more often than not, and it exerts an influence on us that is quite difficult to overcome.

Perhaps it's a combination, then, of some problems with the ability to critically evaluate information along with the compelling nature of social influence that makes cults like Jonestown, or even shared psychotic disorders, possible. But we don't need to look to cults or rare disorders to find examples of strange social influences on our mental lives. Indeed, they're present in every culture throughout the world, including yours.

Penis theft and the power of culture

Take, for instance, the phenomenon of penis theft. Those may not be two words you have seen used together before, but let me assure you: penis theft is a thing.

In 2001, angry mobs in the West African country of Benin killed five people who were suspected of using magic to steal men's penises.[16] The mobs descended on the alleged thieves after men nearby began screaming that their penises had been stolen—in the way a woman in an old movie might yell "Stop, thief!" when a miscreant roughly snags her purse. In Benin, however, the situation took a sharp turn from that old movie scenario and got grisly very fast: the mob doused the suspects in gasoline, lit them on fire, and watched them burn to death—all without a clear indication the suspects had perpetrated *any* theft, much less the theft of someone's genitalia.

It might seem shocking that a mob would be so quick to form—and kill—in response to such an unbelievable accusation. But in some West African regions, penis theft is a constant concern. It's thought to be the result of witchcraft, and, oddly enough, seeing that the penis is still right where it should be doesn't alleviate the intense anxiety a man can experience upon thinking he might have been a victim. These fears of genitalia-focused voodoo also aren't confined to men; there are reports of West African women who claimed someone shrunk their breasts or made their vagina disappear.[17]

It would be easy to brush such beliefs off as silly superstitions. But Western Africa isn't alone in its concerns about penis theft, shrinking, or disappearance. Indeed, these types of beliefs have led to official recognition in diagnostic guides of a disorder known as *koro*, also called *shrinking penis* or *genital retraction syndrome*. Koro mainly occurs in Asian countries, but it's also sometimes seen in various other regions throughout the world. Patients with koro have the unshakeable belief that their penis (or breasts and/or vulva) is shrinking, retracting into the body, or disappearing altogether. Often, there is a fear that the penis retraction, when it finally runs its course, will result in death.

Not surprisingly, these thoughts can lead to panic. They cause physical symptoms such as perspiration and heart palpitations, and often prompt attempts to stop the retraction by pulling on the penis. Sometimes this even becomes a group effort, as family members and neighbors try to help pull the retracting penis out—occasionally tying things like string (or, according to one account from a koro epidemic in China, the stem of a yam)[18] around the penis to pull on it. Not surprisingly, these rescue efforts can result in injuries. In general, though, koro attacks are relatively brief (typically only lasting on the scale of hours) and rarely have any long-term consequences.

Koro is an example of a *culture-bound syndrome*, a condition that is heavily influenced by cultural beliefs and doesn't occur—or at least is interpreted very differently—in other cultures with different belief systems. Thus, by definition, a culture-bound syndrome is dependent on the social dissemination of information. The result is a condition that often sounds absurd to people outside the culture in question.

A better-known example is the evil eye. This phrase has become a slang term for a look of hatred or disgust, but in some cultures throughout the world the evil eye refers to a glance—often spawned from envy or dislike—that carries with it a curse. According to the belief, the curse can cause bad luck, but in some cultures it is considered truly dangerous—potentially leading to injury or death. Belief in the evil eye dates back to the ancient Greeks, but it remains surprisingly widespread in modern times. One study from the 1970s found belief in the evil eye in more than one-third of a sample of 186 societies worldwide.[19]

Probably less familiar to those from Western cultures is the idea of *semen loss anxiety*, also known to the medical community as *Dhat syndrome*. The word *Dhat* comes from Sanskrit and means "elixir of the body" (a reference, of course, to semen). Dhat syndrome is a common condition in several Asian countries, such as India and Pakistan. One study in Pakistan found that 30 percent of men who went to medical clinics reported experiencing Dhat syndrome in the previous month.[20]

The symptoms of Dhat syndrome include fatigue, weakness, anxiety, and sometimes sexual dysfunction, all of which are attributed to the loss of semen. Patients usually claim that most of their semen loss occurs with urination; they often assert they can see semen in their urine, yet the loss of semen is not something doctors—or anyone else—can verify.

Dhat syndrome is linked to a belief in the importance of semen as a "vital fluid" that's essential for health, vigor, and masculinity. Thus, the thought of losing it can cause severe apprehension. Although the origins of Dhat syndrome are unclear, it's frequently attributed to things like excessive masturbation, sexual dreams, overwhelming sexual desire, and the consumption of certain foods, such as eggs or other non-vegetarian foods[21] (Dhat syndrome most often occurs in countries with large vegetarian populations).

Semen-loss anxiety is surprisingly widespread—and common throughout history. Today there are other syndromes that are very similar to Dhat syndrome, such as *shenkui* in China. And semen-loss anxiety is not foreign to Western cultures, either. In the nineteenth century, for example, a condition called *spermatorrhea* was considered a serious public health threat to men in Britain, France, the United States, and a number of other Western countries. Spermatorrhea involved the loss of semen—either involuntarily or through excessive sex or masturbation—and the symptoms mirrored those of Dhat syndrome: weakness, restlessness, fatigue, and so on.[22]

Concerns about issues related to semen loss, and the general thinking that semen should not be wasted (but only used in procreation), helped to develop and bolster arguments that masturbation was a sinful—or at least morally repugnant—practice. Clearly, many cultures think semen is pretty important stuff.

The universality of culture-bound syndromes

It's easy to look at these types of syndromes with a supercilious eye, thinking that these are the beliefs of people who are less advanced than those of your culture. It's important to note, however, that even though these disorders may sound unusual, they are real in the sense that they involve very real symptoms. In other words, even though koro does not truly cause one's penis to disappear, it is associated with extreme anxiety and panic. It's undeniable then, that koro can warrant medical attention (typically in the form of psychiatric help).

And before any readers get too well situated on their high horse, it's also important to emphasize that—even today—no culture is immune to culture-bound syndromes. To every culture, their culture-bound syndromes represent reality, while to others look-ing on, they may seem silly. But you might be surprised to learn that there are common conditions in the modern Western world that some researchers think should be designated as culture-bound syndromes.

Of course, these assertions can be quite controversial, as the mere suggestion that a syndrome is culturally influenced can be considered offensive to those who do experience it. But it's import-ant to note again that if something is designated as a culture-bound syndrome, it doesn't mean it is not real—just that it is heav-ily influenced by culture. Some research, for example, has sug-gested that certain eating disorders, such as *bulimia nervosa*, are culture-bound. In one study of the occurrence of bulimia across different cultures, no evidence was found of the disorder occurring in people who didn't at least have some exposure to (and poten-tial influence from) Western culture and its emphasis on ideals of thinness.[23] Of course this does not mean bulimia is not a legiti-mate disorder, but it suggests that its occurrence may be confined to cultures that have thinness as an ideal—along with enough food availability to promote bingeing.

In some cases, labeling a condition as a culture-bound syn-drome simply means our cultural lens causes us to think about real signs or symptoms in a culturally specific way. Thus, some have proposed (even more controversially) that premenstrual syn-drome (PMS) and attention deficit hyperactivity disorder (ADHD) are candidates for culture-bound syndromes. While both condi-tions involve the undeniable presence of symptoms, whether those symptoms should be considered something that warrants medical treatment varies with cultural perspective. Again, this does not deny that the condition or its symptoms are real. Rather it suggests we may, as a culture, have made decisions to treat the signs and symptoms of menstruation or variations in attention in such a way as to label them as a medical condition, when that decision isn't the only logical conclusion to reach.

Disregarding debates about what should and shouldn't be

considered a culture-bound syndrome, the fact that these syndromes exist underscores the point that social influences have a powerful effect on how we think. It is often said that humans are social creatures by nature, and that statement certainly rings true if we measure its verity by how much value our brains place on social information. Indeed, it seems our brains evolved to value social information so much that our belief in it can occasionally override our better judgment, distort what should be straightforward observations about the world around us—and even alter our typical expectations of what might fall within the confines of reality.

10

ABSENCE

It was a cold and rainy January in London when John had just started what was bound to be one of the more difficult semesters of his college career. He was an undergraduate student working toward his electronics degree, and this semester—in addition to courses on electrical circuit design and analysis—he was taking calculus, general physics, and (much to his dismay as an introverted young man) public speaking.

Midway through the first month of the semester, John was already unsure if he could handle it all. But things were about to get much more complicated. It started when John noticed a funny smell everywhere he went, indoors or out. While it was difficult to pin down exactly what the smell was, it had an earthy pungency that reminded John of rotting fruit. But when he asked his classmates and friends if they smelled it, they said they didn't—and looked at him with a bemused expression that made John feel like he was crazy.

John was experiencing what neurologists call *phantosmia*, or "phantom smell." Something as mild as sinus problems can cause phantosmia, but the symptom can also point to a more serious concern, such as a brain tumor. In John's case, it was the first indication that a herpes infection (yes, that troublesome herpes virus again) had migrated to his brain.

John unsuccessfully tried to ignore the inexplicable smell, but soon he also developed other symptoms: a fever, sore throat, and headache. Maybe, he thought, the strange odor had just been part of getting sick. For the next two days, John felt miserable. His headache and fever got worse. Then his neck became unusually sore.

Although John was unaware of it at the time, his symptoms paint a classic clinical picture of a condition called *meningitis*. Meningitis is an inflammation of the *meninges*—a series of membranes that surround your brain. Those membranes help provide structural support, cushion your brain from potentially damaging blows, and keep the brain from jostling around in the skull. Additionally, they contain a substance called *cerebrospinal fluid*, which not only assists with these protective functions, but also acts as a means to carry important substances throughout the brain and remove waste products from the organ.

The meninges, however, are susceptible to infection from bacteria, viruses, and other pathogens. When one of these germs infiltrates the meninges, it can lead to meningitis—an inflammatory reaction with life-threatening consequences.

It wasn't until John had a seizure that he realized the severity of his situation. Afterward, he was brought to the hospital disoriented and extremely lethargic. Doctors quickly recognized the signs of meningitis, determined that a herpes infection was the root cause, and initiated treatment. After about a week, John's fever subsided and he became more alert.

John's symptoms continued to improve over the next couple of weeks, but he also started to exhibit bizarre behavior during his convalescence: he would try to eat or drink *anything* placed near him. He sipped shampoo, guzzled water from flower vases, and even drank his urine. He tried to eat soap, blankets, his urinary catheter, and his feces. Moreover, he did this all rather nonchalantly, as if it were an everyday occurrence.

Over the next six months, however, John continued to recover, and his unusual eating and drinking behavior gradually disappeared. But he would never get completely back to who he was before his bout with meningitis. He continued to suffer from severe, incapacitating amnesia and displayed extreme mood swings.[1] As bad as all this sounds, I've introduced John's case here to discuss another unusual impairment he developed.

By a couple of months after the meningitis, John was attaining normal scores on tests of verbal IQ and other cognitive function. He spoke fluently, although sometimes he had trouble finding the words he wanted. But when doctors tested his ability to identify the subjects of different pictures by name, they noticed a curious deficit. When they showed John a picture of an inanimate object such as a briefcase, compass, or dustbin, he had no problem saying what the object was or what it was used for. He was, on the other hand, almost unable to identify any living thing.

When doctors specifically tested his ability to recognize living vs. nonliving things, John accurately named 90 percent of the inanimate objects but only 6 percent of living things. When asked to define the word *parrot*, he simply said, "don't know." To *ostrich*, he responded, "unusual." He came a bit closer with a snail when he described it as an "insect animal."[2]

John's deficit was greater than just a linguistic one—he appeared to generally struggle with understanding living things as a classification. This led to an overall inability to properly categorize living things as living, and because this is one of the first steps in identifying what something is, it created difficulties in recognizing things that should have been very familiar to him.

This impairment is unusual for its specificity; how can the brain experience damage that preserves almost all other cognitive capabilities, but does not retain the capacity to identify anything in one particular category? Surprisingly, the specificity of this affliction is not entirely unique. It can be seen in a group of disorders called *agnosias*. Different types of agnosia vary greatly in their overall presentation, but they typically involve the inability to recognize or perceive one distinct class or category of thing.

Known unknowns

Agnosia means "not knowing" in Greek, and the term is used to indicate an impairment in perception or recognition that's not attributable to a sensory or intellectual deficit. John, for example, could still see, and he could think clearly for the most part, but nevertheless he had difficulty recognizing living things. Agnosia itself represents a category of disorders, and within that category is a long list of conditions that often seem quite different from one another on the surface, but all involve the failure to appreciate some aspect of experience that seems fundamental to the rest of us.

Agnosia often emerges after damage to part of the brain that is involved with a particular element of perception. Thus, agnosias help to demonstrate that different parts of your brain are responsible for distinct aspects of perception—ranging from the interpretation of raw sensory data to the integration of such information into a meaningful view of the world. These disparate brain regions must cooperate to make sense of the world around you, and the result enables us to appreciate a perception (e.g., a blooming flower) as well as the contextual importance of it (e.g., an indication that spring is here). However, because disparate brain regions make unique contributions to that complete experience, damage to individual parts of the brain can disrupt our perception of the environment in very particular ways.

We can see this by looking at some different types of agnosia. Patients with *prosopagnosia*, for example, have visual perception that's generally normal—except when it comes to faces. While prosopagnosics can usually tell that a face is a face by identifying features such as a nose, eyes, and so on, they're unable to process the combination of characteristics in such a way as to cause one face to appear distinct from another. In other words, faces to a prosopagnosic are about as distinguishable as knees are to the rest of us.

If, for example, someone with prosopagnosia was walking down the street and their mother came walking the other way, they—in theory—might look their dear old mom directly in the eyes and not recognize her. In reality, this may not happen because most prosopagnosics become extraordinarily adept at using other cues such as hairstyle, clothing, gait, or the sound of one's voice to recognize people—so much so that their symptoms may not be at all apparent to those around them. At the same time, their facial recognition might be so poor that they cannot even identify their own face in the mirror. In especially severe cases of prosopagnosia, facial features may not even be discernable. To one patient, for example, faces "appeared as white globs with two dark circles for eyes" and no other characteristics.[3]

Prosopagnosia is considered a *visual agnosia*, since the impairment primarily involves a deficit in the ability to perceive or identify a visual stimulus (when possible, agnosias are categorized by which sensory modality seems to be most affected). Like other agnosias, visual agnosias are remarkable for the specificity of the stimuli patients cannot perceive properly, which can range from faces to colors to images in the mirror. In the latter case, patients understand they are looking into a mirror—if you asked them, they would confirm it matter-of-factly—but nevertheless they consistently fail to appreciate they are seeing a reflection and not a real three-dimensional scene. They may, for example, repeatedly try to touch items in a mirror, only to be confused when their fingers keep running into the glass.

Visual agnosias, however, can also extend beyond object recognition. Individuals with *akinetopsia*, for instance, are unable to perceive movement. One patient with akinetopsia described how she didn't know when to stop pouring tea because she couldn't see it rising in the cup—as she poured it, the tea seemed frozen in place. She found being near other people disconcerting; because she couldn't register their movements, people appeared to teleport from place to place around her. When she talked to people, their mouths didn't move smoothly but shifted unnaturally from an open to a closed position like a nutcracker doll. Attempting to cross the street was particularly daunting; as she explained, "When I'm looking at the car first, it seems far away. But then...suddenly the car is very near."[4]

Simultanagnosia, a condition where patients can only perceive one object at a time, is just as unsettling. Even when looking at a complex scene with many details, simultanagnosia patients can only focus on one isolated feature. If, for example, you were to sit someone with simultanagnosia down at a fully set dinner table that was laden with several courses of food and ask them what they saw, they might say, "a fork." When looking at a car, they might only see a tire, and when standing in front of a house, a window might be the sole object of their focus. They only see bits and pieces but never the whole. This deficit causes such impairment that simultanagnosia patients are often considered functionally blind.

The nuts and bolts of visual imagery

Studying visual agnosia has helped neuroscientists to appreciate that different aspects of vision are managed by distinct brain regions, because damage to different parts of the brain tends to elicit unique deficits. It is only through the contributions of these individual brain regions (which are part of large networks devoted to visual processing) that your brain is able to piece together a complete visual scene.

Indeed, the segregation of visual information starts in the retina, the thin layer of specialized neurons at the back of the eye where vision begins. The cells in the retina are specialized to respond to light by sending neural signals that initiate the process of visual perception. Most of those neural signals travel to the very back of the brain—to a region called the *primary visual cortex*.

The primary visual cortex receives an overwhelming amount of data. According to one study, the retina sends about 10 million bits of information per second to the brain—comparable to the rate that data travels in a slightly-slower-than-average internet connection.[5] Most of that incoming information is sent to the primary visual cortex, which analyzes it, identifies the basic features of the visual scene—such as

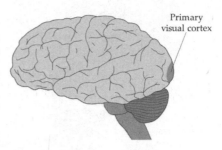

Primary visual cortex

orientation, three-dimensional depth, and direction of movement—and begins reconstructing the scene in the brain.

After your brain has generated a raw image, however, visual processing is far from over. The brain must next assign some meaning to the image. It uses memories of previous experiences to identify familiar aspects of what you are seeing and relies on awareness of current goals to determine which stimuli are most important to focus on. These higher-level features of visual perception involve the recruitment of areas outside the primary visual cortex.

Current neuroscience models suggest that visual information leaves the primary visual cortex and branches off in two directions en route to parts of the brain that are involved in separate aspects of visual processing. This idea is called the *two-streams hypothesis*.

One stream/pathway leaves the primary visual cortex and travels through other surrounding visual areas, which are rather unimaginatively called visual area 2 and visual area 4. The pathway then makes its way to the temporal lobe and stretches down to a region known as the *inferior temporal cortex*, which extends to the underside of the brain. In the inferior temporal cortex, there are columns of neurons that display a specialization for the recognition of specific objects. Thus, this stream—which is called the *ventral stream* because the term *ventral* is used to refer to the lower parts of the brain—helps us to determine what objects are. Because of this, it is also known as the *what pathway*.

The other "stream" leaves the primary visual cortex and travels through visual areas such as visual area 2, visual area 3, and visual area 5 (also known as the *middle temporal visual area*).

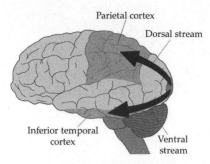

The two streams/pathways that are proposed to exist according to the two-streams hypothesis, and the parts of the brain they extend to. The dorsal stream is sometimes referred to as the where pathway, and the ventral stream is also called the what pathway.

Then, the pathway courses upward toward the parietal cortex. As we discussed in Chapter 2, the parietal cortex has an established role in helping us to understand where our body is in space, and where other objects are in relation to us. Correspondingly, this other visual stream—which is known as the *dorsal stream* because *dorsal* refers to the upper regions of the brain—is thought to be important to identifying the location of things in the environment. Thus, it is often called the *where pathway*.

The anatomical origins of agnosias

Each of these pathways connects numerous parts of the brain, and the result is a complex network that creates rich visual experiences. We can see the unique contributions of these pathways and the regions they connect when we look closer at the anatomical basis of some of the agnosias mentioned above. Patients with prosopagnosia, for example, often have damage to a blueberry-sized region of the inferior temporal cortex that has come to be known as the *fusiform face area*. The fusiform face area contains neurons that are highly active when someone looks at a face—but not when they look at other things such as houses or trees.

Neuroscientists have thus proposed that the fusiform face area is specialized for recognizing faces. This hypothesis (like so many hypotheses) is debated, however, as some argue that the region is attuned to anything we have a high level of familiarity with—not just faces. Indeed, one study found greater activity in the fusiform face area in birdwatchers and car experts when they looked at birds and cars, respectively, and one birdwatcher lost the ability to identify birds after damage to the fusiform face area.[6] However, regardless of the exact role the fusiform face area plays in recognizing faces, the consequences of damage to the region suggest its contributions to face perception are critically important.

In fact, neurons throughout the inferior temporal cortex seem to be arranged in columns that share specializations in identifying particular stimulus characteristics. In most cases, these characteristics are general (e.g., cells might be activated in response to a particular shape and pattern), although neurons such as those in the fusiform face area appear to be exceptions to that rule. In any case, the requirements for activation of cells in the inferior temporal cortex are more complex than a simple size or color, and it's thought that neurons in this region play a critical role in helping us determine what we are looking at. Damage to neurons in the inferior temporal cortex often leads to deficits in recognizing specific visual stimuli (like faces).

Damage to neurons in the *where pathway*, on the other hand, is tied to a disruption of spatial orientation, visual attention, or movement perception. For instance, akinetopsia (the disorder mentioned earlier that causes an impairment in the ability to recognize movement), has been linked to damage to the middle temporal visual area. And simultanagnosia (the disorder that

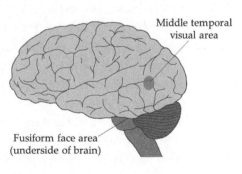

Middle temporal visual area

Fusiform face area
(underside of brain)

causes an inability to perceive more than one object at a time) is associated with damage to areas of the temporal and parietal lobes that are found along the *where pathway.*

Thus, these unusual perceptual disturbances have helped neuroscientists to appreciate the collective neural effort required to accomplish even the simplest visual tasks. But it's not just vision; other types of sensory perception also require the contributions of myriad brain regions, and they can also be disrupted in highly specific ways. In *tactile agnosia*, for example, patients can still use their sense of touch, but they can no longer recognize objects by touch alone. If, for instance, I put a key into your palm and let you feel it with your fingers, it would not take you very long—even with your eyes closed—to realize it is a key. But someone with tactile agnosia might shift the key around in their hand for minutes, running their fingers over every indentation and outline, and still be unable to figure out what it is.

Patients with *amusia* can hear just fine but have trouble perceiving music. They are tone-deaf in the most literal sense of the term. They might be able to identify a song such as "Jingle Bells" due to its recognizable lyrics. But if you took the lyrics out and played an instrumental version, "Jingle Bells" (and any other song) would be a cacophonous jumble of sounds to them. It would be indistinguishable from an instrumental version of "Happy Birthday," "Hey Jude," or "Bohemian Rhapsody." Amusics often find music uninteresting or downright annoying, as to them even the most skillfully played music sounds no better than my fifth grader did in his first couple months of learning the trumpet.

So far, we've focused on agnosias that involve deficits associated with sensory processing—problems related to vision, touch, hearing, and so on. In each case, the impairment disrupts a fractional component of a sensory experience while sparing the rest. But agnosias are not confined to sensory perception. They can also involve a disruption of more complex abilities, leading to the loss of elements of mental life that typically seem indispensable.

Out of time

Gary was a single man in his mid-50s who had followed in his father's footsteps when it came to both his profession (a plasterer) and his chronic drinking. He experienced a fall in the summer of 1938 while intoxicated, and ended up in the hospital experiencing severe confusion. In fact, he was so mentally impaired upon his hospital admission that he couldn't recall his name or birth date. After examining him, doctors found Gary had suffered a skull fracture and had underlying brain damage.

Eventually, Gary was able to answer simple questions, but some of his responses did little to alleviate his doctors' concern. Gary knew who he was and that he was in a hospital. His general reasoning was sound, and his memory was mostly intact. But his most alarming response came when a doctor asked him what year it was. He confidently answered that it was 1895. He was 43 years off the mark; in 1895, Gary had only been 15 years old.

With more questioning, doctors determined that Gary could describe past events well, but he couldn't accurately localize them in time; he knew they had occurred, but he didn't know *when* they had occurred. To him, it seemed just as likely something might have happened yesterday as 10 years ago.

Gary had lost the ability to comprehend, estimate, and appreciate the concept of time. He couldn't guess how long even short intervals of time should feel. If asked to sit quietly for one minute, Gary might say he was done after only 20 seconds—or he might wait for 5 minutes. His deficits were even more apparent when it came to long periods of time, which he tended to shorten drastically. He had gotten divorced 21 years prior, but when asked how long ago he and his wife had separated, he said seven years.[7]

Given that time isn't a sensation, Gary's condition doesn't neatly fall into one of the perceptual categories used to classify other agnosias. So, it has been given its own category and is called *time agnosia*. Time agnosia is difficult to imagine, as for most of us time is an omnipresent consideration. It provides the context for our memories, the urgency for our actions, and an opportunity for the mitigation of our emotional pain. Thankfully, time agnosia is a very rare outcome, typically brought on by brain damage. Its infrequency, however, also acts as an obstacle in coming to understand the neurological basis for the disorder. Neuroscientists have just not had enough patients to study.

Fortunately, Gary's condition improved on its own. In December of 1938—about five months after his hospital admission—he suddenly regained his sense of time. He left the hospital in mid-February of 1939 after a seven-month stay.

A loss of mental imagery

As you read the paragraphs above, you likely formed a series of images in your mind. Maybe you envisioned an older man in a hospital struggling to accurately answer doctors' questions about estimating time. Or perhaps your mind wandered while you were reading, and you conjured up an image of something different altogether. Our minds are constantly creating mental pictures, and the capacity to do so is a pervasive and important part of our lives. We use it to recall details, predict future scenarios, understand new information, and sometimes simply as an escape. But imagine if you lost the capability to form mental images—if the movie projector in your head suddenly went dark. That's what happens to patients with a condition known as *aphantasia*.

Aron was a successful architect in his early 50s when he unexpectedly suffered a stroke. Of course, strokes are never something we really expect, but they are even more of a surprise in someone under 60 and in relatively good health. Fortunately, Aron's age and lack of other serious health conditions likely hastened his recovery, but a few years after his stroke he had some lingering deficits. Aron had developed prosopagnosia severe enough that he had difficulty recognizing his own face in the mirror, and he was easily confused with directions; he had a hard time finding his way around—even in areas he was very familiar with.

But his most striking deficit was the one that would be least detectable to an outside observer: Aron lost his ability to produce mental imagery. This was something he claimed to be adept at before the stroke; it was a skill he used frequently in his career as an architect. "Before, my visualization abilities were pretty impressive," he said. "At my work, I could visualize and remember things that most people had not thought about. I would be sitting there and I would say, well, you can't do X, Y, and Z, because you've got this happening here and there. Now I have to look at the drawing and work my way through it."[8]

Aron couldn't form any pictures in his mind. Not only could he no longer visualize buildings and other structures for his job, but he also could not imagine the faces of friends and family, depictions of recent events, or scenes of places he had been. Even his dreams were total blackness. Fortunately, computers enabled Aron to continue working as an architect, but he was distraught by losing something that was a reliable and comforting feature of his mental life.

No imagination

The first person to conduct a formal study of mental imagery was Francis Galton, a Victorian-era polymath who made important contributions to many fields. Among his notable accomplishments, Galton created the first weather map (the type showing temperatures and cold and warm fronts that is part of every weather forecast today), devised a still-used classification system for fingerprints, and founded the field of behavioral genetics, which explores how genetics influences behavior.*

---

* Unfortunately, Galton's interest in heredity also prompted him to develop the idea of eugenics, a selective breeding approach in humans that Galton

In 1880, Galton conducted a survey of 100 men, many of them distinguished scientists and friends of Galton's. In the survey, Galton asked the respondents to think of an object such as their breakfast table, and to generate an image of that object in their mind. He then asked several questions about the conceived image: Was it hazy or clear? Bright or dim? In color or shades of gray?

To his surprise, most of the survey respondents denied having the ability to clearly form any mental image at all. Galton received a number of responses like this one: "To my consciousness there is almost no association of memory with objective visual impressions. I recollect the breakfast table, but I do not see it." Some were even incredulous that anyone had the capacity for visual imagination. Galton wrote: "To my astonishment, I found that the great majority of men of science...protested that mental imagery was unknown to them...They had no more notion of its true nature than a color-blind man who has not discerned his defect of the nature of color."[9]

However, Galton found that clear mental imagery was common in nonscientists. Thus, Galton confidently declared that "scientific men as a class have feeble powers of visual representation." Galton hypothesized having a scientific mind with proficiency in abstract thought must involve some degree of trade-off, where mental imagery (which Galton thought was more the realm of novelists and poets) is not as well-developed.

Galton's studies of mental imagery went mostly unquestioned in the field of psychology for over 100 years. But eventually others tried to replicate Galton's experiments and failed to find such a large proportion of scientists—or individuals in general—who had difficulty forming mental images. In the early 2000s, a pair of researchers attempted to repeat Galton's survey with modern

saw as a way of improving the human race. Galton's writings on eugenics would become a dark stain on his list of accomplishments, and some would use the idea as a justification for genocide.

scientists and university undergraduates and found that 94 percent of the scientists surveyed indicated a moderate to high capacity for mental imagery. None of them had a lack of mental imagery altogether.[10]

The conclusions Galton drew from his studies have also been questioned. Galton's own research with nonscientists suggested that variation in mental imagery capabilities did not appear to be as specific to scientists as Galton presumed. It's possible Galton's interpretation of the data was biased by his hypotheses, causing him to ignore contradictions to his claims.

Nevertheless, Galton's 1880 "breakfast table study" raised awareness of the possibility that there could be variation in the capacity to generate mental images. There wasn't much more research into this idea until the 2010s, when neurologist Adam Zeman encountered a patient who had lost his ability to produce mental imagery, a deficit that was likely caused by a disruption in blood supply to the brain during a cardiovascular procedure.[11] Previously, the patient claimed to have had a strong capability for mental imagery—something he often used in his work as a surveyor to visualize buildings. Suddenly, however, the pictures in his head had disappeared.

Zeman was intrigued by the case and worked with colleagues to conduct some studies on the patient, whom they referred to as MX. The main goal was to learn more about how the brain had been impacted to produce this unusual deficit, so they used neuroimaging to monitor MX's brain activity while he attempted to generate a mental image. After Zeman and his colleagues published their results, *Discover Magazine* also covered the study in one of their issues. This publicity prompted over 20 people to contact Zeman, claiming to have the same impairment. Zeman's group published another article describing these additional cases; in it they dubbed the condition *aphantasia*, coining the term that roughly translates to "no imagination."[12]

It's uncertain how many people experience aphantasia, although estimates suggest numbers are higher than you might expect. One study, for example, found that over 2 percent of those surveyed claimed to experience "no visual imagination."[13]

Neuroscientists have studied aphantasia patients, as well as people with normal mental imagery, to try to get a better understanding of how the brain produces mental images and what might go wrong to cause those images to vanish. When someone without aphantasia attempts to form a mental image, one area of the brain that is consistently activated is the same region that's involved with normal visual perception: the visual cortex.[14] And when an aphantasia patient tries to generate imagery in their head, their visual cortex is underactive; this is one thing Zeman and his colleagues saw when they looked at MX's brain.[15]

Thus, researchers have suggested that the visual and mental imagery systems share neural components. Damage to these components might disrupt normal visual imagery, mental imagery, or both. In support of this perspective, patients with a specific impairment of visual function also sometimes experience the same type of disruption to their visualization capabilities. One patient, for example, experienced prosopagnosia but also had trouble forming mental images of faces.[16] At the same time, there are patients with visual defects who don't experience mental imagery deficits, and vice versa, which suggests there are unique components to each process as well.

There are a lot of questions still surrounding the neuroscience of aphantasia. Nevertheless, the study of patients with the condition has contributed to our understanding of mental imagery—and helped us to appreciate how important it is in our daily lives.

In this chapter, we've looked at cases where some fundamental aspect of the normal human experience was lost. These cases underscore the many tasks your brain must accomplish to produce the mental life we are all so accustomed to. Even what seem to be the simplest mental functions often require the input of many brain regions, along with healthy networks to connect them.

This arrangement again demonstrates the efficiency—and fragility—of the brain. The way our neurons work together to produce complex experiences such as the awareness of time is nothing short of incredible, but the fact that disrupted neural activity can lead to the loss of such an essential function reminds us that even the most consistent aspects of our experience can be wiped away with alarming ease.

11

DISCONNECTION

L eo was very familiar with the symptoms of a stroke. He had
been the first to notice them and call 911 when his father had
a stroke 10 years earlier. But now, just after being seated for breakfast at a local diner, Leo recognized those same stroke symptoms
again—in himself.

Leo's mind flashed back to his father's stroke. His father had survived but faced an arduous yearlong recovery in which he essentially had to learn how to speak and walk again. The memory of that
demanding experience prompted Leo to consider the long-term
effects he could be facing if he were having a stroke himself. Who
would care for him? He had no children and no partner—would he
have to endure the struggles his father did, but without anyone to
help? Strangely enough, he was more worried about these possibilities than about the immediate risk a stroke posed to his life. And yet,
as Leo monitored his symptoms and apprehensively thought about
what the future had in store, it never once occurred to him that a
stroke would lead to his right hand developing a mind of its own.

Leo's stroke symptoms started when he looked down at the
menu and noticed that he couldn't see anything in his right field
of vision. There was a shadow that covered half the world. It
obscured the words on half of the menu, clouded over the other
tables and patrons situated to his right, and darkened the right
side of the parking lot he should have seen when looking out the
finger-print-smeared restaurant window.

Leo ordered his breakfast, hoping the visual abnormality would be fleeting. But when he went to pick up a cup of coffee, he couldn't even lift it; his hand was incredibly weak. In fact, he realized, it wasn't just his right hand. The whole right side of his body was feeble and nearly paralyzed. At this point, Leo thought, there was no denying that he was having a stroke.

Symptoms of stroke are distinct due to some common characteristics. First, they tend to come on quickly—sometimes within minutes after the brain's blood supply has been disrupted (a disruption in blood supply to the brain is the defining feature of a stroke). Second, typical stroke symptoms—such as numbness, weakness, or visual disturbances—usually affect only one side of the body. This is because the reduction in blood supply that causes a stroke usually impinges (at least initially) on one cerebral hemisphere, and for many functions, one cerebral hemisphere is devoted primarily to the other side of the body.

This is especially true for the functions Leo experienced deficits in: vision and movement. The left side of your brain mainly processes information about the right side of your field of vision, and vice versa. And movement-related signals intended to move the muscles on the right side of the body mostly come from the left side of the brain. Thus, when the workings of one cerebral hemisphere are interfered with, movement and vision on the opposite side of the body are often impaired.

When these types of one-sided symptoms come on quickly and are associated with classic signs of a stroke such as numbness, weakness, trouble speaking, confusion, visual impairment, loss of balance and coordination, and/or severe headache, it paints a pretty clear diagnostic picture.

Leo went to the hospital, where doctors determined he was indeed having a moderately severe stroke. As doctors began treatment, however, they encountered an unexpected obstacle: Leo's right hand.

The first indication of a problem with Leo's hand came when a nurse attempted to hook Leo up to an IV, with the goal of administering drugs to break up the blood clot obstructing blood flow in Leo's brain. As the nurse inserted the needle and began situating the IV lines, Leo's right hand pushed her away, grabbed the IV lines, and started yanking on them.

Leo was mortified by his actions and apologized profusely. He claimed he had no intention of interfering with the nurse's care, and he couldn't explain his behavior. But this was just the start of his hand's insubordination. As doctors continued to treat Leo, his right hand forced itself to be the focus of attention more and more. It would grab the doctor's stethoscope and impede the nurse as she attempted to help. At times it even became violent, trying to slap the medical staff and once taking hold of Leo's own throat and strangling him; Leo needed help to get his hand to release its grip.

Leo could move his arm voluntarily, but he could not prevent it from moving when he didn't want it to, no matter how hard he tried. It seemed to be acting independently of his desires.

Leo spent the next five weeks in the hospital receiving treatment for the lingering effects of his stroke. During that period, he regained the ability to control his right arm as long as he was looking directly at it (when he wasn't, it would resume acting on its own). After being discharged from the hospital, Leo developed other strategies to constrain his unruly hand, such as wearing a heavy weight on his right wrist to keep his arm from lashing out at particularly problematic times. Fortunately, by six months after his discharge, the aberrant activity in Leo's right hand disappeared, and he was able to control it once again.[1]

Alien limbs

The unnerving autonomy displayed by Leo's right hand is typically referred to as *alien hand syndrome*. I say "typically" because, in the hundred-some years since it was first identified, the strange disorder has gone by several names, such as *anarchic hand, le signe de la main étrangère* or "the sign of the foreign hand," and even *Dr. Strangelove syndrome*.*

Alien hand syndrome is rare, but physicians have documented hundreds of cases since it was first described in 1908. Patients with alien hand syndrome share the common characteristic of having a limb (it's usually—but not always—a hand, and some cases of alien leg also occur) that acts with a degree of independence. Sometimes, an alien hand may simply mimic the movements of the patient's opposite hand, but in other cases it is especially ornery—interfering with the patient's actions for no apparent reason. For example, when someone with alien hand syndrome goes to pick something up, the alien hand might knock it back out of his grasp. Or, when he tries to button up his shirt, the alien hand may proceed to unbutton it.

As we saw in Leo's case, occasionally the alien hand's tendencies can be more violent. One woman had to tie her arm down when she slept to keep her hand from trying to choke her in her sleep,[2] and some alien hands also lash out at others, impeding their actions or even outright abusing them. Often, this type of behavior is appalling to the patient. They don't condone the activities of their own limb, and its conduct may be so incongruous with their desires that it feels like it doesn't belong as part of their body—hence justifying the description of it as alien.

The neuroscience of alien hand syndrome has been difficult to explain, in part because it seems there are several different types of dysfunction that can lead to the disorder. The condition typically appears after brain damage—either short-term, such as what occurred with Leo's stroke, or progressive, like what we see in Alzheimer's disease or other conditions that cause the deterioration of the brain. Although there is some variability in the brain regions

---

* Dr. Strangelove is the title character in the Stanley Kubrick film *Dr. Strangelove or: How I Learned to Stop Worrying and Love the Bomb*. Strangelove has a black-gloved hand that acts with a mind of its own; he comically struggles to control it throughout the movie.

that are affected in alien hand patients, many patients have damage to one bundle of nerve fibers in particular: the *corpus callosum*.

The corpus callosum, which we first discussed in Chapter 4, is a large collection of neurons that connects the two cerebral hemispheres. In fact, the 200 million neurons that constitute it make it the largest neuronal pathway in the brain, and it acts as an integral means of communication between the two halves of the organ. The corpus callosum is critical when you consider the divisions of labor between the two cerebral hemispheres. For, when you receive sensory information in one cerebral hemisphere, or when you initiate a movement on one side of your body, it's important that the opposite cerebral hemisphere be made aware of what's going on in the other half of your brain.

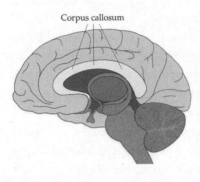

Corpus callosum

For example, when incoming information from your right visual field reaches the left cerebral hemisphere, that information is made available to the whole brain in case you need both hemispheres to coordinate a response to something in the environment. And, when a signal to initiate movement with the right hand originates in the left cerebral hemisphere, your right hemisphere is at the same time made aware of the intention to move—information that might let your right hemisphere know it should step back and let that action proceed without interference.

When the corpus callosum is damaged, this interhemispheric communication is disrupted. In most cases, the brain can adapt to this situation, using alternate pathways to transmit important messages from one hemisphere to the other. In alien hand syndrome patients, however, the interrupted communication may lead to both cerebral hemispheres acting independently to some degree, causing them to fail to coordinate plans for movement.

This lack of harmonization seems to be especially problematic for the nondominant hand (i.e., the left hand in right-handed people), which is the hand that is typically affected in alien hand syndrome.* Neuroscientists believe this is because there are important motor planning areas located in the hemisphere that controls the dominant hand. Without access to the information from those motor planning areas, the nondominant hand cannot properly align its actions with the dominant hand. Thus, the patient ends up like a puppet whose arms are being controlled by two different people, with movements that lack organization and might even appear nonsensical.

This suggested mechanism doesn't explain all the symptoms of alien hand syndrome. It doesn't, for example, account for why an alien hand may become aggressive—behavior that would appear to involve much more than just a lack of coordination with the other limb. Indeed, no hypothesis proposed to explain alien hand syndrome (and there are several) can account for all cases of the disorder, which supports the notion that there may be multiple neurological mechanisms at play. Which mechanism best explains the condition may depend on the patient in question.

In general, however, alien hand syndrome can be held up as an example of how important it is that your two cerebral hemispheres be able to "talk" to one another, and the corpus callosum is the main avenue for that communication to occur. But alien hand syndrome is just one of many consequences that can come about due to the disruption of important connections between different areas of the brain. Other outcomes can be similarly unusual, and they underscore the indispensability of healthy neural communication.

---

* Leo's case was a bit unusual (as far as alien hand syndrome cases go), because his dominant right hand was affected.

At a loss

Ronald was in the grocery store when it became a struggle to move the entire left side of his body. He leaned on the shopping cart to hold himself upright, but he realized he wasn't doing a good job concealing his difficulties when a store employee stopped near him and asked if he was okay. That's when Ronald also discovered he couldn't speak. He wanted to respond to the employee and say, "Yes, I'll be OK," but even though the words were in his mind, he could not get his mouth to voice them.

The store employee called an ambulance, and at the hospital doctors determined Ronald was having a stroke. He survived but faced some substantial impairments early in his recovery. For one, just speaking and understanding speech was an arduous challenge. This deficit alone—the loss of fundamental competencies someone has been relying on since they were a toddler—can be psychologically devastating. Fortunately, after about a month, Ronald's ability to understand language began to return, even though he still struggled to produce it.

As discussed in Chapter 8, these types of language deficits are known as *aphasias*; they're a relatively common outcome of stroke, occurring to some degree in about one-third of stroke survivors.[3] Ronald, however, also experienced a much less common problem. When Ronald had recovered enough to begin feeding himself, he was served a meal in his hospital bed: a piece of turkey, mashed potatoes, and a small bowl of chicken noodle soup. Ronald was hungry, but instead of diving into the meal he sat staring at his dinner tray with an air of uncertainty.

He was looking at the utensils. He knew what these were: fork, spoon, knife. But he could not for the life of him recall how to use them. Eventually, he decided he had to try. He wanted to start with his soup, but which utensil worked best for soup? Hesitatingly, he picked up his knife and began splashing it around in the soup, an action he immediately knew would be ineffectual in getting the soup into his mouth. After several more aborted attempts at feeding himself, Ronald had to call a nurse for help.

Ronald's deficit was not restricted to eating utensils. It soon became apparent that he had lost his understanding of how to use all common tools and objects. When he was shown tools such as a nail clipper or screwdriver, he could say what they were—but he could not use them or even describe how to do so.

During his monthlong stay in the hospital, doctors sometimes provided Ronald with the incorrect tools to complete a task just to see how he responded. When served a meal with a toothbrush, Ronald proceeded to use the toothbrush like a spoon. When it came time to brush his teeth, doctors gave him a spoon instead of a toothbrush (there is a point at which this does seem a bit sadistic). Ronald applied toothpaste to the spoon and began rubbing it across his teeth, unaware of the inappropriate substitution.[4]

As a test, a doctor gave Ronald a coffee mug, water, a spoon, some instant coffee, and an unplugged microwave. Ronald immediately realized his task. "I have to make coffee," he said. But it was soon apparent that Ronald had no clue how to do that. He stared at the objects in front of him as if he were examining an inscrutable foreign text. Then he picked up the plug to the microwave, dangled it into the empty coffee mug, and began twirling it around like he was trying to stir something. He stopped, realizing he had made a mistake but unsure how to correct it.

Finally, Ronald opened the container of instant coffee. He poured a heaping amount into the coffee mug (ignoring the utility of the spoon for this purpose). Then he took the spoon and began stirring the coffee—with no water in the cup. He glanced up at the doctor timidly; it was clear he knew something wasn't right about what he had done. And yet, he could not figure out where he had gone wrong. To his brain, the solution he had arrived at seemed as plausible as any other.[5]

A deficit of action

Ronald was experiencing a condition called *apraxia*, a word that comes from Greek and means "without action." Patients with the disorder develop an impairment in their ability to carry out learned actions. To be diagnosed with apraxia, the difficulty can't be due to some outright loss of motor or sensory function. In other words, Ronald still could move his body in a way that would enable him to make instant coffee, but he couldn't figure out how to go about it.

There are several different types of apraxia. The type Ronald was suffering from is rare; it's known as *conceptual apraxia*. Conceptual apraxia is so named because patients struggle with their conceptual understanding of how to utilize objects like tools to complete tasks—they often cannot recall the most basic details of what a tool is for or how to go about using it. Because of this, they have a difficult time doing things that require tools of any sort. This deficit can, of course, lead to serious problems performing even the simplest functions of daily life.

A more common form of apraxia is *ideomotor apraxia*. In this condition, patients display deficits in performing learned movements despite a lack of underlying motor impairments. This occurs even though the patient understands the movement they want to make. Patients with ideomotor apraxia, for instance, often have a difficult time making hand gestures on command. If you asked someone with ideomotor apraxia to wave goodbye, they would know what they were supposed to do, but their hand probably wouldn't comply. They might just move their hand awkwardly up and down.

Ideomotor apraxia patients can display other deficits with movement as well, including difficulties using tools. But one distinct difference between patients with conceptual apraxia and patients with ideomotor apraxia is that those with ideomotor apraxia understand the movement they want to make, whereas patients with conceptual apraxia are at a loss from the start.

Interestingly, many apraxia patients tend to have more difficulty performing movements when asked to, even if they may be able to perform the movement spontaneously. In other words, a patient with ideomotor apraxia might have no problem naturally waving goodbye to a friend, but if you *asked* them to wave goodbye they would be confused and act as if waving was not something they had yet learned how to do. The reason for this strange discrepancy between their ability to perform spontaneous and requested actions, however, is not clear.

There is a fairly long list of other types of apraxia, each with its own specific deficits. Some focus on certain parts of the body, such as *lid apraxia*, which involves difficulty opening the eyelids despite the muscles needed to do so working just fine. Other forms of apraxia center around specific activities, such as getting dressed (*dressing apraxia*) or speaking (*apraxia of speech*).

The brain in motion

The neuroscience of apraxia varies depending on the type of apraxia in question. Generally, however, apraxia is thought to involve disruptions to brain networks that integrate several things, such as incoming sensory information, a conceptual understanding of a desired action, and a plan for movement. These networks incorporate multiple brain regions, each of which may be critical for a particular type of functional movement to occur.

For example, areas of the parietal cortex are thought to be essential for complex or skilled movements, such as using a screwdriver or sewing a button on. To accomplish such tasks, the parietal lobe receives information from the visual areas of the brain and utilizes that visual feedback to ensure a movement is occurring in the way it is supposed to. Thus, damage to the parietal cortex can have a devastating effect on dexterous movements.[6] But the parietal cortex is also thought to be specifically involved in storing conceptual information about tool use, making it a critically important part of the brain for manipulating objects to serve a specific purpose.[7]

To successfully make any type of complex movement, however, the parietal cortex must communicate with the *motor cortex*, which is found in the frontal lobe. The motor cortex is responsible for initiating most movements and is itself subdivided into several regions that are involved in various aspects of movement, from the planning to the execution of movements.

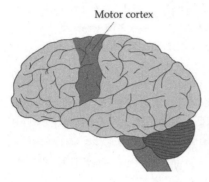

Motor cortex

The connections between the motor cortex and parietal cortex create a basic network that integrates sensory and motor information to enable someone to successfully make the movements necessary to use a tool or another object for a specific purpose. But it probably will come as no surprise at this point that the full network involved in such functions incorporates other regions as well. For example, to do something like use a tool, your brain activation must involve regions that are devoted to formulating goals, selecting the best method for achieving those goals, fine-tuning movements to make them precise, and so on. Thus, damage to any of these areas could produce serious deficits in the ability to perform complex actions. But damage to the pathways that connect the regions can be similarly disruptive, making the connections just as important as the parts of the brain they link together.

The necessity of robust connections in the brain has been an important realization in modern neuroscience. It underscores the significance of brain networks in both healthy brain function and disorders, and reinforces the shift away from focusing solely on individual brain regions in attempting to determine how the brain works. Additionally, a heightened awareness of the importance of brain communication has provided some insight into disorders that consist of such an unusual array of symptoms that it only seems possible to make sense of them by viewing them as the result of disrupted pathways that connect multiple parts of the brain.

A strange collection of symptoms

Sofia was 52 years old when she had a stroke; it left her unable to speak and paralyzed on the right side of her body. Those symptoms, however, disappeared within a few weeks, and soon Sofia appeared to be fully recovered. She was relieved and thought she had been fortunate to make it through the life-threatening experience with no lasting complications. But about a year after her stroke, she developed a new—and highly unusual—collection of symptoms.

It started with headaches, which were often accompanied by nausea. Even though she never had a migraine before, Sofia decided that must be what she was experiencing. Then she also began to feel very unsteady on her feet. When she walked, she did it with a wobbliness that made her appear mildly drunk. She developed some visual problems, especially with seeing things on her right side, and noticed she was having considerable memory lapses. Then—to her bewilderment—she lost her ability to write.

This most recent development prompted Sofia to go back to the hospital. She thought she must be having another stroke—what else could explain these concerning symptoms? Fortunately, the doctors in the hospital found no evidence of a stroke. They did, however, suspect that an inadequate supply of blood to the brain— possibly due to a condition that involves the thickening and narrowing of the brain's arteries—might be responsible for her symptoms.

Her doctors examined her writing ability more closely. Sofia could perform the movements necessary to write, but the resultant writing was an illegible scrawl. Her jumbled letters did not stay on a horizontal line. Instead, they slanted up, down, and in all directions. She was much more successful when copying something already written. Even then, however, she could only duplicate single words accurately; paragraphs inevitably became distorted and full of mistakes as she continued to write.

While examining Sofia, doctors also uncovered some other unusual symptoms. Sofia had lost her comprehension of numbers. Even simple calculations baffled her. Thus, in a matter of weeks, Sofia had gone from being thankful for her complete recovery to losing two functions (writing and the ability to make calculations) that had been dependable aspects of her cognitive life since she was a child.

But that wasn't the end of it. Sofia also displayed a peculiar inability to recognize and differentiate between her own fingers. When asked to use her left hand to touch the middle finger on her right hand, she appeared confused—and somewhat embarrassed. In response, she touched the index finger of her left hand and looked up at the doctor helplessly.

"Do you understand the request?" the doctor asked.

"Yes," Sofia replied.

"Can you repeat the request to me?"

"Use my left hand to touch the middle finger on my right hand. But I—I'm not sure where the middle finger is."

"Can you describe the middle finger to me?"

"Yes, I think … it's longer than the other fingers typically, and … it's in the middle."

"But you cannot pick out your middle finger on your right hand?"

Sofia stared at her hands again and pointed to the small indentation between her ring and middle finger—on her left hand.

As strange as it might seem, Sofia's deficit in identifying fingers is an established (but rare) condition called *finger agnosia*. Recall from Chapter 10 that *agnosia* is a general term used to describe a disorder that involves a deficit in recognition—of any number of things (e.g., faces, living things, movement), depending on the patient's specific impairment. Agnosias are not caused by a sensory deficit; in other words, patients with finger agnosia can still *see*, but they nevertheless are unable to recognize their fingers. Some patients with the condition describe their fingers as an "undifferentiated mass" and claim picking out the individual fingers is nearly impossible.[8]

Sofia's difficulties in tests of finger identification were compounded by another issue: she had lost her understanding of right vs. left. Unlike her fingers, she was able to recognize other parts of her body (such as her ears, eyes, etc.), but when asked to point to one of these on one side of the body her accuracy was no better than a coin flip.

Thus, Sofia's case was made unique by the appearance of four highly unusual symptoms: an inability to write, an inability to perform calculations, finger agnosia, and an impairment in differentiating between right and left. The case I've described is based on an actual patient treated in the 1920s by an Austrian-born neurologist named Josef Gerstmann. Gerstmann recognized the distinctiveness of his patient's symptoms and published a case report about her.[9] A decade later, the collection of symptoms became known as *Gerstmann syndrome*.

Gerstmann suggested that his syndrome might be caused by damage to the parietal lobes.[10] He based this hypothesis on the understanding that the parietal lobe is involved in the perception of one's own body (which seems to be disrupted in finger agnosia). Later neuroscientists have questioned this hypothesis, however, pointing out that there does not seem to be a region in the parietal lobe that could account for all the functions that are disrupted in Gerstmann syndrome.

More recent research suggests that perhaps Gerstmann syndrome is caused by damage to connections that run between multiple brain areas. Although many of those regions are found in the parietal lobe, Gerstmann syndrome seems to be a consequence of disruption to the pathways between brain regions that perform disparate functions, rather than by damage to discrete regions themselves.[11]

Our understanding of Gerstmann syndrome is still evolving. But it, like the other disorders discussed in this chapter, provides an example of the importance of connections in the brain. Without working avenues of communication to transmit information, individual brain regions become isolated, and a brain region that can't communicate with the rest of the brain is essentially nonfunctional. This appreciation of neural communication represents an evolution of understanding in the field of neuroscience—one that may eventually help us to explain some of our strangest neurological disorders.

12

REALITY

Sixty-seven-year-old Olivia was in her kitchen making a cup of tea—a task she didn't expect to be any more memorable than the thousands of other times she had done it in her life. But as she unwrapped a tea bag, she started to experience a strange sensation in her hands. They felt as if they were growing rapidly, and within seconds she had the sense they were five times their normal size. Olivia's seemingly giant hands made unwrapping the comparatively tiny tea bag a nearly impossible task.

Even though Olivia's hands *felt* cartoonishly large, when she looked at them they appeared to be their typical size. This helped reassure her that she was experiencing some sort of perceptual distortion, and she tried to stay calm. She told herself that the strange feeling would fade eventually. After a few minutes, she was proven right, and her hands felt normal again.

The very next day, however, Olivia had another episode where she experienced a peculiar warping of her bodily proportions. This time, she was getting up from a recliner when she felt as if her whole body was growing like an inflating balloon. Olivia instinctively hunched her shoulders and bent her head down for fear she would bump up against the ceiling. Crouched over, she walked to the bathroom to look at herself in the mirror. She wanted to confirm that this episode, like the one she had the day prior, was just in her head.

As she walked, Olivia began to experience an additional—even more disorienting—odd sensation. She had the distinct feeling she was floating upward toward the ceiling.

She stumbled to the bathroom, looked in the mirror, and was reassured to see that she was still normal size. And yet it felt like her gigantic body could barely fit inside the small bathroom. She also was convinced she needed to grip the sink tightly to avoid rising up from the ground and finding herself pressed up against the ceiling. Once again, however, the strange perceptions disappeared after a few minutes.

Over the course of the next week, Olivia experienced several of these episodes. In some, her hands seemed to grow, and in others they shrank. After the episode where she felt her body had become unusually large, she had three others that involved the distinct impression she was miniature in size.

Olivia went to the doctor, who checked her for eye problems, scanned her brain to look for serious issues such as a tumor or a stroke, and ran several other tests; everything came back normal. But Olivia had recently begun taking the antidepressant sertraline (better known as Zoloft), and she confided that when she had taken the same drug 10 years prior, she had experienced similar sensations. After a week, the episodes disappeared, leaving her doctor to assume that Olivia's experiences must have been an exceedingly rare side effect of the antidepressant.[1]

Curiouser and curiouser

The condition Olivia experienced is known as *Alice in Wonderland syndrome*, or AIWS. The reasons for this nomenclature should be obvious to anyone who is even vaguely familiar with Lewis Carroll's popular children's book *Alice's Adventures in Wonderland* (or the movie adaptations of the story). Several times throughout Carroll's book, Alice either grows or shrinks dramatically in size after—rather heedlessly—drinking or eating various unknown substances she finds in a mysterious world at the other end of a rabbit hole.

(Interestingly, some have suggested that Carroll himself experienced AIWS-like symptoms associated with his chronic migraines.)

Like Alice, patients with AIWS often perceive things—both their bodies and other objects in the environment—to be larger or smaller than they really are. But while this is one of the most commonly reported symptoms of AIWS, the syndrome can look very different from case to case, and there have been close to 60 documented symptoms, including: perceiving three-dimensional objects as two-dimensional or flat, seeing everything with a colored tint, seeing multiple copies of one's visual field (almost as if looking through an insect's eye), feeling as if time is sped up or slowed down, and experiencing a sensation of the body being split in two. An episode of AIWS can be quite jarring, to say the least.

Most of these symptoms are not considered hallucinations but instead are referred to as *sensory distortions*. The difference is that hallucinations are generated out of nothing—there is no stimulus to cause them, and they are produced solely in the brain. In a sensory distortion, on the other hand, our perception of something in the environment becomes altered in such a way that it no longer matches reality. It's not unheard of, however, to also experience visual and auditory hallucinations in AIWS.

AIWS is generally considered a rare condition, but its true prevalence is difficult to estimate because there are not even clear diagnostic criteria to define what constitutes an AIWS case. AIWS is, however, surprisingly prevalent in those who experience the relatively common affliction of migraines: the frequency of AIWS among migraine sufferers is thought to be as high as 15 percent.[2] It's not clear what causes the strong link between AIWS and migraines, but AIWS is also associated with a long list of other conditions, such as epilepsy, certain infectious diseases (e.g., Epstein-Barr virus, Lyme disease), stroke, etc.

Failed associations

Parts of the brain that are involved in the processing of visual and/ or somatosensory (i.e., having to do with bodily sensations) information are often implicated in the neuroscience of AIWS. Neuro- imaging studies of AIWS patients suggest symptoms are associated with abnormal activity in the region where the temporal, occipital,

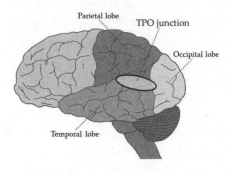

and parietal lobes meet—an area sometimes called the temporo-parieto-occipital, or TPO, junction. This part of the brain is critical for integrating visual informa- tion from the occipital lobe and somatosensory infor- mation from the parietal lobe to create an internal model of ourselves and the outside world. This is essentially an expansion of the body schema discussed in Chapter 2—one that incorporates the virtual representation of the self into a virtual representation of the environment to facilitate our ability to inter- act with the world around us.

The TPO junction is known as an *association area* of the brain, a name that reflects a tendency among neuroscientists to divide the cerebral cortex into three main regions: sensory areas, motor areas, and association areas. The sensory areas are the parts of the brain that receive and process information from the senses, such as touch, smell, taste, and so on. These include regions discussed previously in the book, like the primary visual cortex and primary somatosensory cortex.

The motor areas are, of course, involved in movement. They include the aptly named motor cortex, which we discussed in the last chapter. Motor areas of the brain are responsible for generating movement-related signals to send down to the body, as well as for planning movements and selecting actions based on the goal we're trying to achieve.

Finally, association areas combine sensory and motor informa-tion and use it to understand and interact with the world. While damage to sensory or motor areas can cause predictable deficits in the senses or the ability to produce voluntary movement, respec-tively, damage to association areas tends to lead to more unpre-dictable and complex impairments. We've discussed some of these, such as agnosias and hemispatial neglect, earlier in the book.

Some symptoms of AIWS may thus be caused by a disruption to the association areas of the TPO junction, which integrate sensory information from inside and outside the body. For example, dam-age to parts of the TPO junction that are involved in creating an accurate understanding of the relative relationship (both in terms of size and distance) between ourselves and objects in the envi-ronment may cause symptoms where either our bodies or objects around us appear too large or too small.[3]

Given that there are so many symptoms associated with AIWS, it's likely the explanation of what is going on in the brain to cause it depends a lot on the particular case of AIWS we're dealing with (different cases with different symptoms probably involve dispa-rate brain mechanisms). Due in part to this complexity, there are still a lot of questions about the neurobiology of AIWS. Neverthe-less, AIWS provides a good example of how a disruption to the intricate sensory processing your brain is responsible for can lead to a fantastical experience that aligns better with the content of a children's book than with typical daily life.

The brain's sophisticated hallucinatory capabilities

As I mentioned earlier, AIWS is a sensory distortion, which occurs when the brain incorrectly interprets or fails to integrate some incoming sensory information. But given the right circumstances, the brain is also amazingly adept at creating a perceptual experience where none existed to begin with—a hallucination—which is somewhat surprising given the brain's inclination to strive to create a faithful reconstruction of the environment otherwise. Indeed, your brain can generate a fantastical hallucinatory scene that your eyes would not be able to differentiate from reality, but it rarely does so unless damaged or under the influence of some strong psychedelic drug.

The power of hallucinations is especially striking when they occur in the absence of other sensory phenomena—when, for example, visual hallucinations affect those who otherwise cannot see. This is the case with patients who experience *Charles Bonnet Syndrome*. You might remember Charles Bonnet from Chapter 1; he was the scientist responsible for publishing the first account of what would come to be known as Cotard's syndrome. A few decades before he wrote about the symptoms of Cotard's syndrome, Bonnet documented a hallucinatory condition afflicting his grandfather who had gone blind due to cataracts. Almost 200 years later (in 1967), scientists rediscovered Bonnet's report and—crediting it as the first detailed description of a newly recognized condition—named the syndrome in his honor.

Patients with Charles Bonnet syndrome, or CBS, experience vivid visual hallucinations despite being visually impaired—often severely so. The hallucinations may be simple, such as geometrical shapes materializing in the field of vision or small specks moving across one's view (like an extreme version of the floaters many people experience with age). But CBS hallucinations can also be extraordinarily elaborate.

Truman Abell, a nineteenth-century physician, wrote a highly detailed description of his CBS hallucinations in a letter to the *Boston Medical and Surgical Journal* in 1845.[4] When Truman was 59 years old (in 1838), he began to lose sight in his right eye. Within four years his vision loss had worsened and spread to both eyes, leaving him totally blind. Soon afterward, he began experiencing astonishing visual hallucinations.

In the fall of 1843, Truman was sitting alone by the fireplace when he glanced over and saw a woman sitting several feet away. She was holding a sleeping infant in her arms. After a few minutes, she disappeared, but immediately after she had faded from view, Truman noticed a small child standing near his chair, looking up at him. Truman knew this had to be an illusion, but it looked so real he put out his hand to try to touch the child—only to confirm that there was nothing there.

Over the next year, Truman's hallucinations became more and more complex. By early 1844, he was having elaborate visions of various people and all different kinds of animals. For three weeks straight, he saw a gray horse standing next to him wherever he went. At night, he would see large numbers of people congregating in his room. They frequently came up to his bed and looked down into his face—glaring insolently at him but never speaking.

Often, the walls of Truman's house would fade away, allowing him to look out over hallucinatory open fields illuminated by bright sunlight—even in the middle of the night. One evening at around 10 p.m., Truman visualized a herd of cattle trampling through his house. At first, Truman feared he would be crushed; but he reminded himself that the cattle were not real, and soon they passed by and disappeared.

Once Truman awoke in the night to see hundreds of men, women, and children standing in long columns beginning at the end of his bed and extending far off into the distance. They all seemed to be paying close attention to something—perhaps someone speaking, although Truman couldn't make out any such details. After 15 to 20 minutes, the crowd dispersed and faded into the night.

Another time Truman woke up to find that the walls of his house had disappeared again, revealing an expansive plain with a regiment of soldiers stretching out in two columns as far as the eye could see. The line of soldiers—which seemed to be never-ending—marched past him for hours. The procession continued through the next day and into the following evening. Then, the soldiers turned toward the west and disappeared beyond the horizon. About a week later, Truman again (from his bed) saw countless men, this time on horseback, traveling in a formation he estimated to be a half a mile wide. They continued to file past him for several hours.

While the hallucinations in CBS can be much more mundane than Truman's, his experience is also not extraordinary among those with the condition. Frequently, CBS patients describe seeing people, animals, objects, and so on—sometimes in bizarre relationships with one another (e.g., "snakes crawling out of people's heads"[5]). Typically, though, as fantastical as these visions might be, CBS patients are not distressed by their hallucinations. Instead, they can usually recognize that what they're seeing is not real and wait calmly for it to pass.

Again, a common characteristic of these patients is that they have experienced at least some degree of vision loss; in many cases they are completely blind. And yet, they "see" some of the most outlandish hallucinations imaginable. Indeed, many patients with CBS describe the hallucinations they experience as more vivid and intense than anything they have ever seen with their eyes. How is this possible?

High school biology back to haunt you

Although there are multiple hypotheses to explain CBS, the most popular one centers around the concept of homeostasis. You might remember homeostasis from high school biology—for me, it was one of the few concepts from high school biology that actually stuck (like the axiom that the mitochondrion is the powerhouse of the cell). In biology, homeostasis refers to the proclivity of biological systems to maintain some degree of stability or equilibrium.

Homeostasis is hypothesized to play a role in CBS because when the brain loses visual input that it typically relies on for critical information about the world, it seems to try to find ways to restore the steady level of visual information it has come to expect. One approach it takes is to try to fix the problem by troubleshooting—tinkering in the way you might if you suddenly lost all sound on the video you were watching on your laptop. In that situation, you might turn up the main volume, then (if that didn't do the trick), open up the full volume controls and start fiddling with all the individual volume dials, hoping that manipulating one of them would resolve the issue.

Similarly, the brain turns the dials up or down on different mechanisms to see if that will bring back the visual stimulation it is so reliant on. Some of the adjustments the brain makes cause the cells of the visual system to be more easily excitable. It's as if the brain is increasing the sensitivity of the visual system to encourage more activity there.

This change, however, makes the visual system a little too high-strung. The neurons in the region enter a hyper-aroused state—just waiting to fire—and because of this they are more likely to fire spontaneously, without an external visual stimulus to prompt them to do so. This spontaneous neural activity may cause neurons that would normally be activated by visual stimuli to be activated when no stimulus is actually there. Then, they mistakenly send signals to the rest of the brain indicating the presence of a nonexistent visual stimulus, creating a hallucination.

From out of nothing

This phenomenon of sensory deprivation leading to heightened excitability in the brain, spontaneous signaling, and hallucinations does not just occur in CBS. In fact, one of the surest ways to induce hallucinations is to deprive someone of their normal sensory input. Prisoners kept in isolation in the dark (or blindfolded), for example, often have vivid hallucinations.[6] And sensory deprivation tanks—lightproof, soundproof chambers constructed to allow for as little sensory stimulation as possible—have long been touted (and sought out by the more psychedelically inclined) as a way to elicit hallucinations.[7]

Researchers have even used sensory deprivation to evoke hallucinations. In one 2004 study, for example, experimenters blindfolded a small group of 13 participants 24 hours a day for 5 days to see if the experience might induce visual hallucinations. It did, in 10 of the 13 participants. The hallucinations they experienced ranged in complexity from simple flashes of light to elaborate images that resembled the hallucinations of a CBS patient. For example, one of the participants in the experiment said, "I had an image of an older woman with a very wrinkled face...she was facing me. She seemed to be sitting in an airplane seat....And then the scene changed from a woman's face to a mouse-like face, not necessarily smaller but with the features of a mouse."[8]

In some participants, the hallucinations took several days to begin, but in others they appeared just a few hours after blindfolding. The potential for hallucinations to emerge so rapidly may support another hypothesis on how sensory deprivation can lead to visual hallucinations. According to this perspective, spontaneous activity (i.e., activity that is generated by your brain and not associated with a stimulus in the environment) in the visual cortex is common, but normally that spontaneous activity is drowned out by the constant flow of real visual information we receive from our eyes. When that real visual input is cut off, it causes the spontaneous activity to lead to visual signals—and resultant hallucinations. This hypothesis is not mutually exclusive with the homeostatic imbalance hypothesis discussed earlier; different hallucinatory experiences might involve slightly different—or multiple—mechanisms.

In any case, one common feature of hallucinations is that they tend to be linked to activity in the part of the brain devoted to the types of perceptions that occur during the hallucination. For example, when researchers took CBS patients and put them into a neuroimaging device that measures brain activity, the parts of the brain that are typically active when seeing images similar to what was experienced in the hallucinations were the parts that lit up.[9] Specifically, hallucinations of faces were associated with activity in the fusiform face area (the part of the brain devoted to processing visual information about faces that we discussed in Chapter 10), colorful hallucinations were linked to areas of the brain devoted to color processing, and so on.

Thus, visual hallucinations tend to be linked to activity in the brain that is very similar to the activity that occurs when we are actually seeing something, but hallucinations don't involve activity in the retina or the pathways that carry visual information from the eye to the brain. In hallucinations, then, the neural signaling is all centralized within the brain. In other ways, however, there seems to be little difference between a visual hallucination and visual perception. The same can be said for hallucinations experienced in other senses, such as hearing, touch, and so on.

One interesting note about CBS hallucinations is that, while it's not uncommon for patients with CBS to hallucinate images of other people, the people they visualize are rarely familiar to them. Instead, the hallucinations tend to take on a depersonalized quality, and they almost never interact with the hallucinator beyond facial expressions or gestures. This type of indifference is in stark contrast to the hallucinations I'll discuss next, which are of a deeply personal nature. Indeed, perhaps nothing could be more personal.

Seeing ghosts

Samotracia was 70 years old when she started working with a new psychiatrist, Dr. Carlos Sluzki, at a clinic for low-income patients in California.[10] Samotracia had been seeing another therapist for two years but had made little progress. Her previous therapist had diagnosed Samotracia with schizophrenia, but her symptoms did not respond to the drugs typically given to schizophrenic patients, and therapy hadn't been productive. She was hoping a new therapist—especially one who was bilingual and had a multicultural background, like Dr. Sluzki—would be a change for the better.

Through their initial counseling sessions, Dr. Sluzki learned the details of Samotracia's life. She had been born in Mexico and entered the United States illegally with her husband after getting married at an early age. She and her husband had four children (two sons and two daughters), but eventually separated due to her husband's alcoholism and associated violent behavior. Samotracia had lived a difficult life, working full-time while also raising her four children.

Now, at age 70, she had good relationships with her daughters. She talked to them frequently over the phone and occasionally visited them in person. Her sons, however, had both died years ago. One had been killed in gang-related violence when he was a teenager; the other had died of AIDS at age 33.

After their first several appointments, Dr. Sluzki was confused about why Samotracia's previous therapist had diagnosed her with schizophrenia. Schizophrenia can involve an assortment of symptoms ranging from a lack of emotional expression to delusions and hallucinations, but so far Samotracia had displayed none of the characteristic signs of the disorder. When the discussion turned to her sons, though, Dr. Sluzki recognized the basis for the diagnosis.

Samotracia confided that, for the past several years, she had been visited by her deceased sons three to four times a week. The visits would typically occur in the evening, while Samotracia was relaxing after dinner. When they visited, her sons had normal conversations with Samotracia. They talked and joked with her and reassured her that she didn't need to worry about them—they were all right.

In general, these evening visits were almost indistinguishable from what a visit with her sons might have been like if they were still living. Her boys "monkeyed around" a bit, but they also were caring and respectful. They even demonstrated proper manners, letting Samotracia have privacy if she wanted to do something such as shower or get changed. They only visited her at home and didn't appear in other less private settings. And they seemed completely real. As Samotracia told Dr. Sluzki, "Doctor, most of the times I can see and hear them as clearly as I can see you."

Samotracia said she thought it was unlikely the visits were supernatural in nature. Instead, she claimed they were probably a product of her imagination. At the same time, she wasn't sure—and she didn't dwell on it too much for fear that questioning the experience would somehow affect the frequency of the visits. The last thing she wanted was to risk doing something that would cause the visits to end. She felt like the opportunity to see her sons regularly was a gift—a way for her to maintain relationships with all her children, something she had craved for years since the death of her sons.

Dr. Sluzki viewed Samotracia's experiences with her sons as a healthy coping mechanism for dealing with her losses. Because the visions of her dead sons were the only symptom she had that would align with the criteria for schizophrenia (other symptoms would be necessary for an accurate schizophrenia diagnosis), Dr. Sluzki decided that Samotracia wasn't schizophrenic. He treated Samotracia's experiences with her sons as hallucinations, but also didn't seek out methods to stop the hallucinations from occurring. When Samotracia moved to a new area after working with Dr. Sluzki for over a year, she was still seeing her deceased sons on a regular basis.

Bereavement hallucinations

How you interpret Samotracia's experiences might depend on your cultural background and personal beliefs. Of course, alleged encounters with spirits of the deceased are nothing new—we can find references to ghosts going back to ancient Mesopotamia. Even through modern times, ghosts have remained a common part of almost every culture throughout the world. A 2021 survey of Americans, for example, found that 20 percent of them say they have had an encounter with a ghost, and 41 percent believe that ghosts exist.[11]*

---

* For what it's worth, 11 percent of the sample also recalled experiences with demons and 11 percent also claimed to have had an encounter with other

During the past 50 years or so, researchers have become more interested in these types of occurrences—not so much as true experiences of the supernatural, but as grief-induced disturbances in perception, or *bereavement hallucinations*. The use of the term hallucination in this context is a bit controversial, as it carries with it something of a pathological implication, when in fact many mental health professionals recognize that these types of experiences may help with the grieving process. Additionally, many people who have these experiences believe they are representative of true interactions with the deceased, something that the use of the term hallucination contradicts.

Because these types of encounters are often highly personal and meaningful for those who have them, I do not want to dismiss them or diminish their importance by trying to explain them away with a scientific perspective. At the same time, scientists avoid hypotheses that posit the involvement of the supernatural, because such hypotheses are difficult (or impossible) to test. Thus, when attempting to explain an event such as an encounter with a deceased loved one, scientists will opt for the explanation that involves a known, testable, and reproducible mechanism—like a hallucination—rather than an explanation that involves something poorly understood, not testable, and not reproducible, such as a visitor from the afterlife. Of course, if convincing evidence of interaction with spirits is obtained, then scientists will have to revise their paradigms. But until then, science will assume there is not a place for the supernatural in understanding our daily lives. Thus, I will use the term *bereavement hallucination* because it concisely reflects how the scientific community currently views these types of episodes.

Research suggests that bereavement hallucinations are fairly

---

supernatural beings. Four percent said they had a personal encounter with a werewolf, and 3 percent had an experience with a vampire; the margin of error for the sample, however, was 4 percent, which somewhat negates these last two results.

common. According to an analysis of 21 studies, over half of people who had lost a family member or friend experienced a bereavement hallucination of some sort. The nature of the experiences varied. The most common was simply a sense that the deceased person was nearby—this occurred in almost 40 percent of the hallucinations. Over 22 percent of those who experienced bereavement hallucinations also talked with the deceased, and more than 20 percent saw them. Sometimes they just heard the deceased's voice, but in other cases they could smell or touch them.[12]*

Studies have found bereavement hallucinations to be even more prevalent among widows and widowers. One study conducted with residents of nursing homes, for example, suggested that 61 percent of widows had experienced the presence of their husband after his death. Seventy-nine percent of these experiences involved the widows seeing their spouse, and 18 percent involved talking with him.[13] Researchers hypothesize that this increased prevalence of bereavement hallucinations among widows and widowers is due simply to the strong, long-term attachments these individuals often had with the deceased; breaking such an attachment causes more stress and an increased likelihood for the brain to respond in a way that generates a hallucinatory experience.

Neuroscientists, of course, are interested in what might be happening in the brain during bereavement hallucinations. One hypothesis that has been offered up to explain the occurrence of bereavement hallucinations involves a mechanism of the brain known as *predictive coding*. According to this perspective, one of the most important functions of the brain—indeed, depending on whom you ask, *the* most important function—is its role in using past information combined with what's going on in the moment to predict what we should expect to experience next. This

---

* It's worth noting that this study was retracted, but it was retracted for plagiarism, not for any issue with methodology. Thus, the results still seem to be valid. I cited it simply because there are not many studies on the topic to choose from, and this is the most comprehensive recent study available.

is happening constantly, and many neuroscientists think it's the basis for sensory perception.

When you, for example, look around the environment you are in, you are bombarded with sensory information. It is, indeed, an onslaught of data. As mentioned in an earlier chapter, one estimate suggests that the retina alone sends about 10 million bits of information per second to your brain.[14] And this is just from the visual system; your brain must be prepared to handle comparable levels of signaling from other sensory systems as well: hearing, smell, touch, taste, and more.* Your brain has to process all that raw sensory data to make sense of it—and it must do so nearly instantaneously if it expects to be able to use that information to survive in potentially life-threatening situations (which is something it evolved to do).

To be efficient in these endeavors, your brain makes guesses based on the information it has at hand. If you are home in your room, for instance, you might see your jacket hanging from your door and your brain would predict it is just that: a jacket. But imagine instead that you are staying at your aunt's old house that smells funny and always gives you the creeps. You get up in the middle of the night to get a drink of water, and as you're tiptoeing across the cold, creaky wooden floor in the moonlit darkness, you catch sight of yourself in a mirror—and see a large shadowy figure standing behind you. You freeze. Your heart immediately begins to race, and you apprehensively turn your head to get a better look at the ominous shadow, thinking this must be the ghost of your dead uncle or some other poltergeist that has been haunting this musty old house for years. When you look behind you, you find... it's just a jacket hanging from the door.

The same stimulus in both cases caused very different effects

---

* The "and more" refers to the fact that scientists believe we have more senses than the traditional five senses. Proprioception, for example, is the sense of where your body is in space, equilibrioception is the sense of balance, nociception is the sense of pain, and so on.

because your brain wasn't just perceiving them as they were, it was predicting what they might be based on previous experience and current information about your body and emotional state. It took the cues from your heightened vigilance in your aunt's house and thought, "there might be something dangerous here—be on guard." And so, it predicted an ambiguous stimulus could be a potentially menacing one. Of course, it adjusted that prediction once the prediction was determined to be inaccurate.

According to the predictive coding perspective, the sensory experiences we have are primarily predictions that are eventually confirmed by experience—not true representations of what is going on around us. In other words, your brain is constructing your perception of the world based on its predictions, not on what is actually there. It's only when reality doesn't mesh with your brain's prediction that the brain adjusts and causes you to have a more direct experience with the real world.

In this sense, *all* of your experience is a hallucination—albeit one that's controlled by your brain. But in the case of actual hallucinations, such as the bereavement hallucinations we've been discussing, it may be that the brain's predictions overpower its inclination to make corrections. In other words, the brain might give more weight to the predictions than it does to reality, resulting in the predictions, in a sense, becoming reality.

Why would the brain err in this way? Well, it could be influenced by many factors, but one clear culprit would be the stress of losing a loved one. This overwhelming strain might make the brain more prone to the mistaken overvaluation of predictions. Additionally, the brain might be more inclined to predict the presence of an individual whose presence it had come to expect over a period of years. In other words, if a widow's husband had often been in the house with her for 50 years, then it might make more sense for the brain to erroneously predict his presence in the house—even after his death. And an individual's beliefs—cultural, religious, and spiritual—would play a large role in determining how likely they are to either posit the presence of a lost loved one in the first place (for example, if someone were less open to the possibility of such experiences, it might make the brain less likely to predict they would occur).

At the same time, these are just hypotheses. Bereavement hallucinations are difficult to study, as they often occur at unpredictable times and aren't amenable to being reproduced in the laboratory. Thus, we don't know for sure what's going on in the brain when someone experiences a bereavement hallucination. And until we know that, then who's to say that some of these cases don't involve true interactions with the afterlife?

CONCLUSION

Psychiatry has traditionally taken an all-or-nothing approach to disorders: either a patient has one or they don't, and that determination is made based on a set of clear diagnostic criteria. But there is a growing appreciation among scientists for the idea that almost any type of behavior can be considered part of a spectrum of human tendencies—with one end of that spectrum representing an excess of the behavior, and the other end indicating its absence.

Of course, behavior that is dominated by either of these extremes may be problematic, but those who spend more time in the middle of the spectrum are not free of abnormal propensities—we all display them now and then. Someone, for example, who does not have obsessive-compulsive disorder (OCD) may still occasionally experience obsessive thoughts, and they might engage in compulsive behavior every now and again. The difference for someone with OCD is in the degree and frequency, which is what causes the condition to become a major obstacle in their life (and a diagnosable disorder).

The same can be said for so many of the behaviors discussed in this book. Despite how strange some of them may seem, they often just represent the extremes of the spectrum of normal human tendencies—and they are not completely foreign to us. Many of us, for example, treat objects we spend a lot of time with as if they have their own personalities (have you ever gotten angry at a piece of technology as if it were intentionally malfunctioning just to spite you?), have misperceptions about how our body really looks, or act like different people in different situations. But as we've seen, the atypical amplification of these ordinary human characteristics can become pathological, overwhelming, and highly distressing. Still, the fact that we are familiar with the normal behavior that lies at the root of the pathology can make even these unusual cases somewhat relatable.

At the other end of the spectrum, some of the cases we've discussed involve the distinct loss of an ability we expect to be perpetually present. Nevertheless, our familiarity with the faculties that are lost causes us to be able to appreciate the experience of these individuals as well. We can recognize the impact an impairment in the ability to read, speak, or perceive faces would have because these are capabilities we rely on every day.

What I'm trying to say is, despite how peculiar many of the behaviors discussed in this book may seem, they are all occurring in people whose brains are not very different from yours or mine. Indeed, some of the tendencies we've covered (such as the heavy reliance on social information) are things that all of us display. Other behaviors may only emerge when something goes awry, but they involve the exaggeration or underutilization of an aspect of the normal human experience. And, as I've mentioned repeatedly, none of us are immune to circumstances that can cause the type of drastic neurological changes we've seen in the preceding chapters. So, despite the title, I encourage you to think of the behaviors described in this book not as bizarre exceptions, but as examples of the full span of the human condition. They are undeniably part of the human behavioral repertoire.

If your brain works the way it should, then congratulations. Appreciate this moment while it lasts. Because your brain—and the rest of your body—will not work this well forever. Our brains are magnificent organic machines, but like all machines, they will eventually fail. So, take advantage of the functionality of your brain while you still can: create memories, experience emotion, indulge in pleasure (and practice restraint), think deeply, engage your body—do all those things that your brain permits you to do, and do them with great gusto. Never take your brain, and the capabilities it endows you with, for granted.

But perhaps just as importantly, practice compassion for those whose brains work differently. Because truthfully, all of us have some "abnormalities" in our brain function. For many of us, we become expert at hiding these anomalous thought patterns away—keeping them clandestine at all costs. But the more you study the brain, the more you realize that the idea of a "normal" brain is a bit unrealistic, at least in the way we typically conceptualize it. We are all imperfect, and we are all sometimes plagued by thoughts and perceptions that hinder our happiness, sanity, and well-being. Just embracing and being open about this reality, as well as accepting it in others, would go a long way to improving the collective mental health of our society.

ACKNOWLEDGMENTS

Something nobody tells you when you decide to write a book is just how darn difficult it is. Of course, this was my second book, so I learned that lesson the first time around. *Bizarre*, however, involved some unexpected challenges. I did much of my writing in the midst of the COVID-19 pandemic, while also fulfilling my professional responsibilities to The Pennsylvania State University—often at home where I was quarantined with my family for long stretches of time. I sometimes found myself reflecting with bitter irony on the fact that I was writing a book focused heavily on disorders of the human psyche, while the stress of my work and the unpredictability of world events made me feel less and less psychologically stable. Without my family, I'm not sure how I would have found the resiliency to finish this project. So, my first "thank you" goes to them.

I cannot thank my wife Michelle enough—for so much more than I could ever summarize in an acknowledgments page of a book. Thank you for coming to understand my eccentricities and—instead of running away as fast as you could—learning to deal with them in an understanding and caring way. And thank you for continuing to support my ventures, even when they might seem ill-advised or poorly timed. I owe much of what I've accomplished in the past decade to you being there to help me achieve it.

Ky and Fia, although sometimes you were part of the reason I felt unhinged over the past couple of years, you were also responsible for lifting my spirits so many times it's impossible to count. I wouldn't have picked anyone else to be quarantined with while I was writing this book. I can only hope that one day it makes you half as proud to be able to pick up this book and say, "My dad wrote this," as it makes me every day to watch you grow into amazing human beings.

And of course, thank you to my wonderfully supportive parents. Without your help and patience while I figured out what I wanted to do with my life, writing books about neuroscience would be just a daydream.

I'm grateful to my agent, Linda Konner, who once again believed in my idea when it was uncertain if anyone else would develop the same faith in it. I hope that sticking by my side will one day end up being a lucrative decision for you.

Thank you to the team at Nicholas Brealey Publishing for helping me to put my ideas to print for a second time. Thanks to Jonathan Shipley for seeing the potential in the book and to Brett Halbleib for your editorial savvy. And thank you once again to Michelle Surianello; your expertise, professionalism, and attention to detail have been a bright point in the publication process for both of my books.

I am incredibly appreciative of everyone who took the time to peruse my manuscript and offer feedback, advice, and endorsements. Thanks to Tom Gould for reading through the book in its earlier stages and providing valuable and insightful suggestions. And thanks to Bill Ray, Kate Anderson, Kristen Breit, Erin Kirschmann, Alison Kreisler, Amy Stading, and Allison Wilck for reading the manuscript and letting me know what you thought. I still (even the second time around) find the generosity that goes into such an act to be surprising and profoundly admirable.

Finally, thank you to all the individuals whose stories I have told in this book. Many of those stories involve great suffering and distress, and although in most cases I have adapted the details from other sources (and not gotten them directly from the patients themselves), I still feel the need to express my gratitude to be able to retell them. I know it's unlikely my statement of appreciation will be seen by those whose stories I've told herein, but I hope I have portrayed their lives in a way that they would find sensitive, accurate, and valuable.

ENDNOTES

Introduction

1 G.M. Lavergne, *A Sniper in the Tower: The Charles Whitman Murders* (Denton, Texas: University of Texas Press, 1997), 82.

Chapter 1

1 J.L. Saver, "Time is brain—quantified," *Stroke* 37, no. 1 (January 2006): 263–66.

2 H. Förstl and B. Beats, "Charles Bonnet's description of Cotard's delusion and reduplicative paramnesia in an elderly patient (1788)," *British Journal of Psychiatry* 160, no. 3 (March 1992): 416–18.

3 S. Dieguez, "Cotard Syndrome," *Frontiers of Neurology and Neuroscience* 42 (2018): 23–34.

4 Ibid.

5 A.W. Young and K.M. Leafhead, "Betwixt life and death: case studies of Cotard delusion," in *Method and Madness: Case Studies in Neuropsychiatry*, ed. P.W. Halligan and J.C. Marshall (East Sussex, England: Taylor & Francis, 1996), 147–71.

6 H. Debruyne, M. Portzky, F. Van den Eynde, and K. Audenaert, "Cotard's syndrome: a review," *Current Psychiatry Reports* 11, no. 3 (June 2009): 197–202.

7 A.W. Young and K.M. Leafhead, "Betwixt life and death: case studies of Cotard delusion," 147–71.

8 P. Johansson, L. Hall, S. Sikström, and A. Olsson, "Failure to detect mismatches between intention and outcome in a simple decision task," *Science* 310, no. 5745 (October 2005): 116–19.

9 E.C. Hunter, M. Sierra, and A.S. Alex, "The epidemiology of depersonalisation and derealisation. A systematic review," *Social Psychiatry and Psychiatric Epidemiology* 39, no. 1 (January 2004): 9–18.

10 M.P. Alexander, D.T. Stuss, and D.F. Benson, "Capgras syndrome: a reduplicative phenomenon," *Neurology* 29, no. 3 (March 1979): 334–39.

11 C. Pandis, N. Agrawal, and N. Poole, "Capgras' delusion: a systematic review of 255 published cases," *Psychopathology* 52, no. 3 (July 2019): 161–73.

12 V.S. Ramachandran, "Consciousness and body image: lessons from phantom limbs, Capgras syndrome and pain asymbolia," *Philosophical Transactions of the Royal Society of London B: Biological Sciences* 353, no. 1377 (November 1998): 1851–59.

13 W. Hirstein and V.S. Ramachandran, "Capgras syndrome: a novel probe for understanding the neural representation of the identity and familiarity of persons," *Proceedings of the Royal Society of London B: Biological Sciences* 264, no. 1380 (March 1997): 437–44.

14 H.D. Ellis, "The role of the right hemisphere in the Capgras delusion," *Psychopathology* 27, no. 3-5 (1994): 177–85.

15 K.W. de Pauw, T.K. Szulecka, and T.L. Poltock, "Frégoli syndrome after cerebral infarction," *The Journal of Nervous and Mental Disease* 175, no. 7 (July 1987): 433–38.

16 R.J. Berson, "Capgras' syndrome," *American Journal of Psychiatry* 140, no. 8 (August 1983): 969–78.

17 J.L. Mulcare, S.E. Nicolson, V.S. Bisen, and S.O. Sostre, "The mirror sign: a reflection of cognitive decline?" *Psychosomatics* 53, no. 2 (March–April 2012): 188–92.

18 A. Villarejo, V.P. Martin, T. Moreno-Ramos, A. Camacho-Salas, J. Porta-Etessam, and F. Bermejo-Pareja, "Mirrored-self misidentification in a patient without dementia: evidence for right hemispheric and bifrontal damage," *Neurocase* 17, no. 3 (June 2011): 276–84.

19 M.F. Shanks and A. Venneri, "The emergence of delusional companions in Alzheimer's disease: an unusual misidentification syndrome," *Cognitive Neuropsychiatry* 7, no. 4 (November 2002): 317–28.

Chapter 2

1 P.E. Keck, H.G. Pope, J.I. Hudson, S.L. McElroy, and A.R. Kulick, "Lycanthropy: alive and well in the twentieth century," *Psychological Medicine* 18, no. 1 (February 1988): 113–20.

2 J.D. Blom, "When doctors cry wolf: a systematic review of the literature on clinical lycanthropy," *History of Psychiatry* 25, no. 1 (March 2014): 87–102.

3 Ibid.

4 R.B. Khalil, P. Dahdah, S. Richa, and D.A. Kahn, "Lycanthropy as a culture-bound syndrome: a case report and review of the literature," *Journal of Psychiatric Practice* 18, no. 1 (January 2012): 51–4.

5 A.G. Nejad and K. Toofani, "Co-existence of lycanthropy and Cotard's syndrome in a single case," *Acta Psychiatrica Scandinavica* 111, no. 3 (March 2005): 250–52.

6 K. Rao, B.N. Gangadhar, and N. Janakiramiah, "Lycanthropy in depression: two case reports," *Psychopathology* 32, no. 4 (July 1999): 169–72.

7 Ibid.

8 M. Benezech, J. De Witte, J.J. Etcheparre, and M. Bourgeois, "A lycanthropic murderer," *American Journal of Psychiatry* 146, no. 7 (July 1989): 942.

9   H. Flor, L. Nikolajsen, and T.S. Jensen, "Phantom limb pain: a case of mal-adaptive CNS plasticity?" *Nature Reviews Neuroscience* 7, no. 11 (November 2006): 873–81.

10 S.R. Weeks, V.C. Anderson-Barnes, and J.W. Tsao, "Phantom limb pain: theories and therapies," *Neurologist* 16, no. 5 (September 2010): 277–86.

11 V.S. Ramachandran and W. Hirstein, "The perception of phantom limbs. The D.O. Hebb lecture," *Brain* 121, no. 9 (September 1998): 1603–30.

12 M.T. Padovani, M.R. Martins, A. Venâncio, and J.E. Forni, "Anxiety, depression and quality of life in individuals with phantom limb pain," *Acta Ortopédica Brasileira* 23, no. 2 (March–April 2015): 107–10.

13 P.W. Halligan, J.C. Marshall, and D.T. Wade, "Unilateral somatoparaphrenia after right hemisphere stroke: a case description," *Cortex* 31, no. 1 (March 1995): 173–82.

14 T.E. Feinberg, A. Venneri, A.M. Simone, Y. Fan, and G. Northoff, "The neuroanatomy of asomatognosia and somatoparaphrenia," *Journal of Neurology, Neurosurgery and Psychiatry* 81, no. 3 (March 2010): 276–81.

15 H.O. Karnath and C. Rorden, "The anatomy of spatial neglect," *Neuropsychologia* 50, no. 6 (May 2012): 1010–17.

16 Ibid.

17 P.M. Jenkinson, N.M. Edelstyn, J.L. Drakeford, C. Roffe, and S.J. Ellis, "The role of reality monitoring in anosognosia for hemiplegia," *Behavioural Neurology* 23, no. 4 (2010): 241–43.

18 J. Money, R. Jobaris, and G. Furth, "Apotemnophilia: Two cases of self-demand amputation as paraphilia," *The Journal of Sex Research* 13, no. 2 (May 1977): 115–25.

19 P.D. McGeoch, D. Brang, T. Song, R.R. Lee, M. Huang, and V.S. Ramachandran, "Xenomelia: a new right parietal lobe syndrome," *Journal of Neurology, Neurosurgery, and Psychiatry* 82, no. 12 (December 2011): 1314–19.

20 C. Dyer, "Surgeon amputated healthy legs," *BMJ* 320, no. 7231 (February 2000): 332.

21 E.D. Sorene, C. Heras-Palou, and F.D. Burke, "Self-amputation of a healthy hand: a case of body integrity identity disorder," *The Journal of Hand Surgery: British & European Volume* 31, no. 6 (December 2006): 593–95.

22 B.D. Berger, J.A. Lehrmann, G. Larson, L. Alverno, and C.I. Tsao, "Nonpsychotic, nonparaphilic self-amputation and the internet," *Comprehensive Psychiatry* 46, no. 5 (September–October 2005): 380–83.

Chapter 3

1   I. Yurdaisik, "Role of radiology in pica syndrome: a case report," *Eurasian Journal of Critical Care* 3, no. 1 (2021): 33–5.

2   B.E. Johnson and R.L. Stephens, "Geomelophagia. An unusual pica in iron-deficiency anemia," *American Journal of Medicine* 73, no. 6 (December 1982): 931–32.

3 E.O. Bernardo, R.I. Matos, T. Dawood, and S.L. Whiteway, "Maternal cauto-pyreiophagia as a rare cause of neonatal hemolysis: a case report," *Pediatrics* 135, no. 3 (March 2015): e726–29.

4 C.M. Meier and R. Furtwaengler, "Trichophagia: Rapunzel syndrome in a 7-year-old girl," *The Journal of Pediatrics* 166, no. 2 (February 2015): 497.

5 E.P. Lacey, "Broadening the perspective of pica: literature review," *Public Health Reports* 105, no. 1 (January–February 1990): 29–35.

6 A.K.C. Leung and K.L. Hon, "Pica: a common condition that is commonly missed: an update review," *Current Pediatrics Reviews* 15, no. 3 (2019): 164–69.

7 E.J. Fawcett, J.M. Fawcett, and D. Mazmanian, "A meta-analysis of the worldwide prevalence of pica during pregnancy and the postpartum period," *International Journal of Gynecology & Obstetrics* 133, no. 3 (June 2016): 277–83.

8 D.E. Danford and A.M. Huber, "Pica among mentally retarded adults," *American Journal of Mental Deficiency* 87, no. 2 (September 1982): 141–46.

9 C. Borgna-Pignatti and S. Zanella, "Pica as a manifestation of iron deficiency," *Expert Review of Hematology* 9, no. 11 (November 2016): 1075–80.

10 D.E. Vermeer and D.A. Frate, "Geophagia in rural Mississippi: environmental and cultural contexts and nutritional implications," *The American Journal of Clinical Nutrition* 32, no. 10 (October 1979): 2129–35.

11 R.K. Grigsby, B.A. Thyer, R.J. Waller, and G.A. Johnston Jr., "Chalk eating in middle Georgia: a culture-bound syndrome of pica?," *Southern Medical Journal* 92, no. 2 (February 1999): 190–92.

12 *Eat White Dirt*. Directed by A. Forrester. Wilson Center for Humanities and Arts, 2015. adamforrester.com/eat-white-dirt.

13 S. Hergüner, I. Ozyildirim, and C. Tanidir, "Is Pica an eating disorder or an obsessive-compulsive spectrum disorder?," *Progress in Neuro-Psychopharmacology and Biological Psychiatry* 32, no. 8 (December 2008): 2010–11.

14 "Obsessive-Compulsive Disorder," National Institute of Mental Health, accessed May 25, 2022, https://www.nimh.nih.gov/health/statistics/obsessive-compulsive-disorder-ocd.

15 S. Rachman, "Fear of contamination," *Behaviour Research and Therapy* 42, no. 11 (November 2004): 1227–55.

16 C.N. Burkhart and C.G. Burkhart, "Assessment of frequency, transmission, and genitourinary complications of enterobiasis (pinworms)," *International Journal of Dermatology* 44, no. 10 (October 2005): 837–40.

17 M.L. Nguyen, M.A. Shapiro MA, and S.J. Welch, "A case of severe adolescent obsessive-compulsive disorder treated with inpatient hospitalization, risperidone and sertraline," *Journal of Behavioral Addictions* 1, no. 2 (June 2012): 78–82.

18 T. McBride, S.E. Arnold, and R.C. Gur, "A comparative volumetric analysis of the prefrontal cortex in human and baboon MRI," *Brain, Behavior and Evolution* 54, no. 3 (September 1999): 159–66.

19 R.O. Frost, D.F. Tolin, G. Steketee, K.E. Fitch, and A. Selbo-Bruns, "Excessive acquisition in hoarding," *Journal of Anxiety Disorders* 23, no. 5 (June 2009): 632–39.

20 K. Mack, "Alone and buried by possessions," *Chicago Tribune*, August 10, 2010, https://www.chicagotribune.com/news/ct-xpm-2010-08-10-ct-met-hoarders -0811-20100810-story.html.

21 "People v. Suzanna Savedra Youngblood," Animal Legal & Historical Center, accessed May 30, 2022, https://www.animallaw.info/case/people-v-youngblood.

22 "Kyrstal R. Allen, Appellant, v. Municipality of Anchorage, Appellee," Animal Legal & Historical Center, accessed May 30, 2022, https://www.animal law.info/case/allen-v-municipality-anchorage.

23 T.E. Morrisseey, "*Hoarders*: Man Lives with Thousands of Uncaged Pet Rats," *Jezebel*, January 11, 2011, https://jezebel.com/hoarders-man-lives-with -thousands-of-uncaged-pet-rats-5730682.

24 D.F. Tolin and A. Villavicencio, "Inattention, but not OCD, predicts the core features of hoarding disorder," *Behaviour Research and Therapy* 49, no. 2 (February 2011): 120–25.

25 J.R. Grisham, R.O. Frost, G. Steketee, H.J. Kim, and S. Hood, "Age of onset of compulsive hoarding," *Journal of Anxiety Disorders* 20, no. 5 (2006): 675–86.

26 C.M. Hough, T.L. Luks, K. Lai, O. Vigil, S. Guillory, A. Nongpiur, S.M. Fekri, et al., "Comparison of brain activation patterns during executive function tasks in hoarding disorder and non-hoarding OCD," *Psychiatry Research: Neuroimaging* 255 (September 2016): 50–9.

27 E. Volle, R. Beato, R. Levy, and B. Dubois, "Forced collectionism after orbito-frontal damage," *Neurology* 58, no. 3 (February 2002): 488–90.

Chapter 4

1 F. Peek, *The Real Rain Man: Kim Peek* (Utah: Harkness Publishing Consultants, 1996), 7.

2 D.A. Treffert and D.D. Christensen, "Inside the Mind of a Savant," *Scientific American* 293, no. 6 (December 2005): 108–13.

3 F. Peek, *The Real Rain Man: Kim Peek*, 9.

4 A.W. Snyder, E. Mulcahy, J.L. Taylor, D.J. Mitchell, P. Sachdev, and S.C. Gandevia, "Savant-like skills exposed in normal people by suppressing the left fronto-temporal lobe," *Journal of Integrative Neuroscience* 2, no. 2 (December 2003): 149-58.

5 Ibid.

6 A. Snyder, H. Bahramali, T. Hawker, and D.J. Mitchell, "Savant-like numerosity skills revealed in normal people by magnetic pulses," *Perception* 35, no. 6 (2006): 837-45.

7 J. Gallate, R. Chi, S. Ellwood, and A. Snyder, "Reducing false memories by magnetic pulse stimulation," *Neuroscience Letters* 449, no. 3 (January 2009): 151-54.

8 T. Rehagen, "Uncharted Waters," *Southwest Magazine*, October 2016, 56-77.

9 D.A. Treffert and D.L. Rebedew, "The savant syndrome registry: a preliminary report," *WMJ* 114, no. 4 (August 2015): 158–62.
10 S. Keating, "The Violent Attack that Turned a Man into a Maths Genius," BBC, July 8, 2020, https://www.bbc.com/future/article/20190411-the-violent -attack-that-turned-a-man-into-a-maths-genius.
11 D.A. Treffert, "The sudden savant: a new form of extraordinary abilities," *WMJ* 120, no. 1 (April 2021): 69–73.
12 D.A. Treffert, "Brain Gain: A Person Can Instantly Blossom into a Savant— and No One Knows Why," *Scientific American,* July 25, 2018, https://blogs .scientificamerican.com/observations/brain-gain-a-person-can-instantly -blossom-into-a-savant-and-no-one-knows-why/.
13 Ibid.

Chapter 5

1 J. Simner, J.E.A. Hughes, and N. Sagiv, "Objectum sexuality: a sexual orientation linked with autism and synaesthesia," *Scientific Reports* 9, no. 1 (December 2019): 19874.
2 *Horizon*, "Derek Tastes of Earwax." *Daily Motion*, 48:54. September 30, 2004. https://www.dailymotion.com/video/x1olkn1.
3 J. Simner, J.E.A. Hughes, and N. Sagiv, "Objectum sexuality," 19874.
4 C. Scorolli, S. Ghirlanda, M. Enquist, S. Zattoni, and E.A. Jannini, "Relative prevalence of different fetishes," *International Journal of Impotence Research* 19, no. 4 (July–August 2007): 432–37.
5 S. Freud, "Fetishism," trans. J. Strachey, in *The Complete Psychological Works of Sigmund Freud* (London: Hogarth and the Institute of Psychoanalysis, 1976), 147–57.
6 J.J. Plaud and J.R. Martini, "The respondent conditioning of male sexual arousal," *Behavior Modification* 23, no. 2 (February 1999): 254–68.
7 V.S. Ramachandran, "Phantom limbs, neglect syndromes, repressed memories, and Freudian psychology," *International Review of Neurobiology*, 37 (1994): 291–333
8 W. Mitchell, M.A. Falconer, and D. Hill, "Epilepsy with fetishism relieved by temporal lobectomy," *Lancet* 267, no. 6839 (September 1954): 626–30.
9 Ibid.
10 R. Dwaraja and J. Money, "Transcultural sexology: formicophilia, a newly named paraphilia in a young Buddhist male," *Journal of Sex & Marital Therapy* 12, no. 2 (1986): 139–45.
11 L. Shaffer and J. Penn, "A Comprehensive Paraphilia Classification System," in *Sex Crimes and Paraphilia*, ed. E.W. Hickey (Upper Saddle River, New Jersey: Pearson/Prentice Hall, 2006), 69–93.
12 S.S. Boureghda, W. Retz, F. Philipp-Wiegmann, and M. Rösler, "A case report of necrophilia—a psychopathological view," *Journal of Forensic and Legal Medicine* 18, no. 6 (August 201): 280–84.

13 E. Ehrlich, M.A. Rothschild, F. Pluisch, and V. Schneider, "An extreme case of necrophilia," *Legal Medicine (Tokyo)* 2, no. 4 (December 2000): 224–26.

14 C.C. Joyal and J. Carpentier, "The prevalence of paraphilic interests and behaviors in the general population: a provincial survey," *The Journal of Sex Research* 54, no. 2 (February 2017): 161–71.

15 J. Drescher, "Out of DSM: Depathologizing Homosexuality," *Behavioral Sciences (Basel)* 5, no. 4 (December 2015): 565–75.

16 F.J. Jiménez-Jiménez, Y. Sayed, M.A. García-Soldevilla, and B. Barcenilla, "Possible zoophilia associated with dopaminergic therapy in Parkinson disease," *Annals of Pharmacotherapy* 36, no. 7-8 (July-August 2002): 1178–79.

17 A.H. Evans and A.J. Lees, "Dopamine dysregulation syndrome in Parkinson's disease," *Current Opinion in Neurology* 17, no. 4 (August 2004): 393–98.

18 J.M. Burns and R.H. Swerdlow, "Right orbitofrontal tumor with pedophilia symptom and constructional apraxia sign," *Archives of Neurology* 60, no. 3 (March 2003): 437–40.

Chapter 6

1 Baer R., *Switching Time: A Doctor's Harrowing Story of Treating a Woman with 17 Personalities* (New York: Three Rivers Press, 2007), 91.

2 O. van der Hart, R. Lierens, and J. Goodwin, "Jeanne Fery: a sixteenth-century case of dissociative identity disorder," *Journal of Psychohistory* 24, no. 1 (Summer 1996): 18–35.

3 H. Faure, J. Kersten, D. Koopman, and O. van der Hart, "The 19th century DID case of Louis Vivet: new findings and re-evaluation," *Dissociation* 10, no. 2 (1997): 104–13.

4 Reuters, "Tapes Raise New Doubts About 'Sybil' Personalities," *The New York Times*, August 19, 1998, https://www.nytimes.com/1998/08/19/us/tapes-raise-new-doubts-about-sybil-personalities.html.

5 R.W. Rieber, "Hypnosis, false memory and multiple personality: a trinity of affinity," *History of Psychiatry* 10, no. 37 (March 1999): 3–11.

6 D. Nathan., *Sybil Exposed: The Extraordinary Story Behind the Famous Multiple Personality Case* (New York: Free Press, 2011).

7 E.M. Vissia, M.E. Giesen, S. Chalavi, E.R. Nijenhuis, N. Draijer, B.L. Brand, and A.A. Reinders, "Is it trauma- or fantasy-based? Comparing dissociative identity disorder, post-traumatic stress disorder, simulators, and controls," *Acta Psychiatrica Scandinavica* 134, no. 2 (August 2016): 111–28.

8 M.J. Dorahy, B.L. Brand, V. Sar, C. Krüger, P. Stavropoulos, A. Martínez-Taboas, R. Lewis-Fernández, and W. Middleton, "Dissociative identity disorder: an empirical overview," *Australian & New Zealand Journal of Psychiatry* 48, no. 5 (May 2014): 402–17.

9 H. Strasburger and B. Waldvogel, "Sight and blindness in the same person: gating in the visual system," *Psych Journal* 4, no. 4 (December 2015): 178–85.

10 A.A. Reinders, E.R. Nijenhuis, A.M. Paans, J. Korf, A.T. Willemsen, and J.A. den Boer, "One brain, two selves," *Neuroimage* 20, no. 4 (December 2003): 2119–25.

11 C.J. Dalenberg, B.L. Brand, D.H. Gleaves, M.J. Dorahy, R.J. Loewenstein, E. Cardeña, P.A. Frewen, et al., "Evaluation of the evidence for the trauma and fantasy models of dissociation," *Psychological Bulletin* 138, no. 3 (May 2012): 550–88.

12 A.A. Nicholson, M. Densmore, P.A. Frewen, J. Théberge, R.W. Neufeld, M.C. McKinnon, and R.A. Lanius, "The dissociative subtype of posttraumatic stress disorder: Unique resting state functional connectivity of basolateral and centromedial amygdala complexes," *Neuropsychopharmacology* 40, no. 10 (September 2015): 2317–26.

13 C.W. Berman, "Out of His Body: A Case of Depersonalization Disorder," *HuffPost*, September 11, 2011, https://www.huffpost.com/entry/depersonali zation-disorder_b_953909.

14 T.A. Clouden, "Dissociative amnesia and dissociative fugue in a 20-year-old woman with schizoaffective disorder and post-traumatic stress disorder," *Cureus* 12, no. 5 (May 2020): e8289.

15 P. Sharma, M. Guirguis, J. Nelson, and T. McMahon, "A case of dissociative amnesia with dissociative fugue treatment with psychotherapy," *Primary Care Companion for CNS Disorders* 17, no. 3 (May 2015).

16 N. Medford, M. Sierra, D. Baker, and A.S. David, "Understanding and treating depersonalisation disorder," *Advances in Psychiatric Treatment* 11, no. 2 (2005): 92–100.

17 A. Staniloiu and H.J. Markowitsch, "Dissociative amnesia," *Lancet Psychiatry* 1, no. 3 (August 2014): 226–41.

18 D. Sakarya, C. Gunes, E. Ozturk, and V. Sar, "'Vampirism' in a case of dissociative identity disorder and post-traumatic stress disorder," *Psychotherapy and Psychosomatics* 81, no. 5 (2012): 322–3.

Chapter 7

1 C.K. Meador, "Hex death: voodoo magic or persuasion?," *Southern Medical Journal* 85, no. 3 (March 1992): 244–47.

2 "Cancer Stat Facts: Esophageal Cancer," Surveillance, Epidemiology, and End Results (SEER) Program, National Cancer Institute, accessed May 30, 2022, https://seer.cancer.gov/statfacts/html/esoph.html.

3 J.K. Boitnott, "Clinicopathologic conference. Case presentation," *The Johns Hopkins Medical Journal* 120, no. 3 (1967): 186–99.

4 Ibid.

5 W.B. Cannon, "Voodoo death," *Psychosomatic Medicine* 19, no. 3 (May-June 1957): 182–90.

6 A.J. de Craen, T.J. Kaptchuk, J.G. Tijssen, and J. Kleijnen, "Placebos and placebo effects in medicine: historical overview," *Journal of the Royal Society of Medicine* 92, no. 10 (October 1999): 511–15.

7   F. Benedetti, "Beecher as Clinical Investigator: Pain and the Placebo Effect," *Perspectives in Biology and Medicine* 59, no. 1 (2016): 37–45.

8   H.K. Beecher, "The powerful placebo," *Journal of the American Medical Association* 159, no. 17 (December 1955): 1602–6.

9   J.D. Levine, N.C. Gordon, and H.L. Fields, "The mechanism of placebo analgesia," *Lancet* 2, no. 8091 (September 1978): 654–57.

10 R. de la Fuente-Fernández, T.J. Ruth, V. Sossi, M. Schulzer, D.B. Calne, and A.J. Stoessl, "Expectation and dopamine release: mechanism of the placebo effect in Parkinson's disease," *Science* 293, no. 5532 (August 2001): 1164–66.

11 A.J. Crum, W.R. Corbin, K.D. Brownell, and P. Salovey, "Mind over milkshakes: mindsets, not just nutrients, determine ghrelin response," *Health Psychology Journal* 30, no. 4 (July 2011): 424–29.

12 M.U. Goebel, A.E. Trebst, J. Steiner, Y.F. Xie, M.S. Exton, S. Frede, A.E. Canbay, et al., "Behavioral conditioning of immunosuppression is possible in humans," *The FASEB Journal* 16, no. 14 (December 2002): 1869–73.

13 W.S. Agras, M. Horne, and C.B. Taylor, "Expectation and the blood-pressure-lowering effects of relaxation," *Psychosomatic Medicine* 44, no. 4 (September 1982): 389–95.

14 K. Meissner, "Effects of placebo interventions on gastric motility and general autonomic activity," *Journal of Psychosomatic Research* 66, no. 5 (May 2009): 391–98.

15 C. Butler and A. Steptoe, "Placebo responses: an experimental study of psychophysiological processes in asthmatic volunteers," *British Journal of Clinical Psychology* 25, pt. 3 (September 1986): 173–83.

16 C.J. Beedie, E.M. Stuart, D.A. Coleman, and A.J. Foad, "Placebo effects of caffeine on cycling performance," *Medicine & Science in Sports & Exercise* 38, no. 12 (December 2006): 2159–64.

17 M. Darragh, B. Yow, A. Kieser, R.J. Booth, R.R. Kydd, and N.S. Consedine, "A take-home placebo treatment can reduce stress, anxiety and symptoms of depression in a non-patient population," *Australian & New Zealand Journal of Psychiatry* 50, no. 9 (September 2016): 858–65.

18 E. Rogev and G. Pillar, "Placebo for a single night improves sleep in patients with objective insomnia," *The Israel Medical Association Journal* 15, no. 8 (August 2013): 434–38.

19 S.J. Lookatch, H.C. Fivecoat, and T.M. Moore, "Neuropsychological Effects of Placebo Stimulants in College Students," *Journal of Psychoactive Drugs* 49, no. 5 (November–December 2017): 398–407.

20 V. Hoffmann, M. Lanz, J. Mackert, T. Müller, M. Tschöp, and K. Meissner, "Effects of placebo interventions on subjective and objective markers of appetite: a randomized controlled trial," *Frontiers in Psychiatry* 18, no. 9 (December 2018): 706.

21 I. Kirsch, B.J. Deacon, T.B. Huedo-Medina, A. Scoboria, T.J. Moore, and B.T. Johnson, "Initial severity and antidepressant benefits: a meta-analysis of data

submitted to the Food and Drug Administration," *PLoS Medicine* 5, no. 2 (February 2008): e45.

22 C.G. Goetz, J. Wuu, M.P. McDermott, C.H. Adler, S. Fahn, C.R. Freed, R.A. Hauser, et al. "Placebo response in Parkinson's disease: comparisons among 11 trials covering medical and surgical interventions," *Movement Disorders* 23, no. 5 (April 2008): 690–99.

23 S. Rheims, M. Cucherat, A. Arzimanoglou, and P. Ryvlin, "Greater response to placebo in children than in adults: a systematic review and meta-analysis in drug-resistant partial epilepsy," *PLoS Medicine* 5, no. 8 (August 2008): e166.

24 J.B. Moseley, K. O'Malley, N.J. Petersen, T.J. Menke, B.A. Brody, D.H. Kuykendall, J.C. Hollingsworth, et al., "A controlled trial of arthroscopic surgery for osteoarthritis of the knee," *New England Journal of Medicine* 347, no. 2 (July 2002): 81–88.

25 R. Sihvonen, M. Paavola, A. Malmivaara, A. Itälä, A. Joukainen, H. Nurmi, J. Kalske, et al.; Finnish Degenerative Meniscal Lesion Study (FIDELITY) Group, "Arthroscopic partial meniscectomy versus sham surgery for a degenerative meniscal tear," *New England Journal of Medicine* 369, no. 26 (December 2013): 2515–24.

26 W. Häuser, E. Hansen, and P. Enck, "Nocebo phenomena in medicine: their relevance in everyday clinical practice," *Deutsches Ärzteblatt International* 109, no. 26 (June 2012): 459–65.

27 N. Mondaini, P. Gontero, G. Giubilei, G. Lombardi, T. Cai, A. Gavazzi, and R. Bartoletti, "Finasteride 5 mg and sexual side effects: how many of these are related to a nocebo phenomenon?," *The Journal of Sexual Medicine* 4, no. 6 (November 2007): 1708-12.

28 G. Makris, N. Papageorgiou, D. Panagopoulos, and K.G. Brubakk, "A diagnostic challenge in an unresponsive refugee child improving with neurosurgery-a case report," *Oxford Medical Case Reports* 2021, no. 5 (May 2021): 161–64.

29 T.J. Snijders, F.E. de Leeuw, U.M. Klumpers, L.J. Kappelle, and J. van Gijn, "Prevalence and predictors of unexplained neurological symptoms in an academic neurology outpatient clinic—an observational study," *Journal of Neurology* 251, no. 1 (January 2004): 66–71.

30 J. Stone, A. Carson, R. Duncan, R. Roberts, C. Warlow, C. Hibberd, R. Coleman, et al., "Who is referred to neurology clinics?—the diagnoses made in 3781 new patients," *Clinical Neurology and Neurosurgery* 112, no. 9 (November 2010): 747–51.

31 V. Voon, C. Gallea, N. Hattori, M. Bruno, V. Ekanayake, and M. Hallett, "The involuntary nature of conversion disorder," *Neurology* 74, no. 3 (January 2010): 223–38.

32 D.L. Drane, N. Fani, M. Hallett, S.S. Khalsa, D.L. Perez, and N.A. Roberts, "A framework for understanding the pathophysiology of functional neurological disorder," *CNS Spectrums* (September 2020): 1–7.

Chapter 8

1 B. Sharma, R. Handa, S. Prakash, K. Nagpal, I. Bhana, P.K. Gupta, S. Kumar, et al., "Posterior cerebral artery stroke presenting as alexia without agraphia," *The American Journal of Emergency Medicine* 32, no. 12 (December 2014): 1553.e3-4.

2 B.D. McCandliss, L. Cohen, and S. Dehaene, "The visual word form area: expertise for reading in the fusiform gyrus," *Trends in Cognitive Sciences* 7, no. 7 (July 2003): 293–99.

3 B. Okuda, K. Kawabata, H. Tachibana, M. Sugita, and H. Tanaka, "Postencephalitic pure anomic aphasia: 2-year follow-up," *Journal of the Neurological Sciences* 15, no. 187(1-2) (June 2001): 99–102.

4 "Massive Proportion of World's Population are Living with Herpes Infection," World Health Organization, last modified May 1, 2020, https://www .who.int/news/item/01-05-2020-massive-proportion-world-population -living-with-herpes-infection.

5 H. Damasio, "Neuroanatomical Correlates of the Aphasias," in *Acquired Aphasia*, ed. M. Sarno (New York: Academic Press, 1998), 43–68.

6 R.D. Freeman, S.H. Zinner, K.R. Müller-Vahl, D.K. Fast, L.J. Burd, Y. Kano, A. Rothenberger, et al., "Coprophenomena in Tourette syndrome," *Developmental Medicine and Child Neurology* 51, no. 3 (March 2009): 218–27.

7 A. Yamadori, E. Mori, M. Tabuchi, Y. Kudo, and Y. Mitani, "Hypergraphia: a right hemisphere syndrome," *Journal of Neurology, Neurosurgery and Psychiatry* 49, no. 10 (October 1986): 1160–64.

8 S. Finger, "Paul Broca (1824–1880)," *Journal of Neurology* 251, no. 6 (June 2004): 769–70.

9 E.D. Ross, "The aprosodias. Functional-anatomic organization of the affective components of language in the right hemisphere," *Archives of Neurology* 38, no. 9 (September 1981): 561–69.

10 A.K. Lindell, "In your right mind: right hemisphere contributions to language processing and production," *Neuropsychology Review* 16, no. 3 (September 2006): 131–48.

11 J. Greenhalgh, "A Curious Case of Foreign Accent Syndrome," NPR, June 1, 2011, https://www.npr.org/sections/health-shots/2011/06/01/136824428/a -curious-case-of-foreign-accent-syndrome.

12 S. Keulen, J. Verhoeven, E. De Witte, L. De Page, R. Bastiaanse, and P. Mariën, "Foreign accent syndrome as a psychogenic disorder: a review," *Frontiers in Human Neuroscience* 10 (April 2016): 168.

13 P. Mariën, S. Keulen, and J. Verhoeven, "Neurological aspects of foreign accent syndrome in stroke patients," *Journal of Communication Disorders* 77 (January–February 2019): 94–113.

14 Y. Higashiyama, T. Hamada, A. Saito, K. Morihara, M. Okamoto, K. Kimura, H. Joki, et al., "Neural mechanisms of foreign accent syndrome: lesion and network analysis," *Neuroimage: Clinical* 31 (2021): 102760.

15 L. McWhirter, N. Miller, C. Campbell, I. Hoeritzauer, A. Lawton, A. Carson, and J. Stone, "Understanding foreign accent syndrome," *Journal of Neurology, Neurosurgery and Psychiatry* 90, no. 11 (November 2019): 1265–69.

Chapter 9

1 K. Dewhurst and J. Todd, "The psychosis of a association; folie à deux," *The Journal of Nervous and Mental Disease* 124, no. 5 (November 1956): 451–59.
2 P. Wehmeier, N. Barth, and H. Remschmidt, "Induced delusional disorder. a review of the concept and an unusual case of folie à famille," *Psychopathology* 36, no. 1 (January–February 2003): 37–45.
3 E. Daniel and T.N. Srinivasan, "Folie à famille: delusional parasitosis affecting all the members of a family," *Indian Journal of Dermatology, Venereology and Leprology* 70, no. 5 (September-October 2004): 296–67.
4 E. Asp, K. Manzel, B. Koestner, C.A. Cole, N.L. Denburg, and D. Tranel, "A neuropsychological test of belief and doubt: damage to ventromedial prefrontal cortex increases credulity for misleading advertising," *Frontiers in Neuroscience* 6 (July 2012): 100.
5 M.P. Jensen, G.A. Jamieson, A. Lutz, G. Mazzoni, W.J. McGeown, E.L. Santarcangelo, A. Demertzi, et al., "New directions in hypnosis research: strategies for advancing the cognitive and clinical neuroscience of hypnosis," *Neuroscience of Consciousness* 3, no. 1 (2017): 1–14.
6 E. Facco, C. Bacci, and G. Zanette, "Hypnosis as sole anesthesia for oral surgery: the egg of Columbus," *Journal of the American Dental Association* 152, no. 9 (September 2021): 756–62.
7 D. Bruno, "Hypnotherapy Isn't Magic, But it Helps Some Patients Cope with Surgery and Recovery," *The Washington Post*, November 9, 2019, https://www.washingtonpost.com/health/hypnotherapy-as-an-alternative-to-anesthesia-some-patients--and-doctors--say-yes/2019/11/08/046bc1d2-e53f-11e9-b403-f738899982d2_story.html.
8 Z. Dienes and S. Hutton, "Understanding hypnosis metacognitively: rTMS applied to left DLPFC increases hypnotic suggestibility," *Cortex* 49, no. 2 (February 2013): 386–92.
9 Ibid.
10 M. Kilduff and P. Tracey, "Inside Peoples Temple," *New West Magazine*, August 1, 1977, https://jonestown.sdsu.edu/?page_id=14025.
11 E. Asp, K. Ramchandran, and D. Tranel, "Authoritarianism, religious fundamentalism, and the human prefrontal cortex," *Neuropsychology* 26, no. 4 (July 2012): 414–21.
12 K.E. Croft, M.C. Duff, C.K. Kovach, S.W. Anderson, R. Adolphs, and D. Tranel, "Detestable or marvelous? Neuroanatomical correlates of character judgments," *Neuropsychologia* 48, no. 6 (May 2010): 1789–801.
13 J.M. Curtis and M.J. Curtis, "Factors related to susceptibility and recruitment by cults," *Psychological Reports* 73, no. 2 (October 1993): 451–60.

14   L.L. Dawson, "Who joins new religious movements and why: twenty years of research and what have we learned?," *Studies in Religion/Sciences Religieuses* 25, no. 2 (1996): 141–61.

15   S.E. Asch, "Studies of independence and conformity: I. A minority of one against a unanimous majority," *Psychological Monographs: General and Applied* 70, no. 9 (1956): 1–70.

16   "Benin Alert over 'Penis Theft' Panic," *BBC News*, November 27, 2001, World, http://news.bbc.co.uk/2/hi/africa/1678996.stm.

17   V.A. Dzokoto and G. Adams, "Understanding genital-shrinking epidemics in West Africa: koro, juju, or mass psychogenic illness?" *Culture, Medicine, and Psychiatry* 29, no. 1 (March 2005): 53–78.

18   W.S. Tseng, K.M. Mo, J. Hsu, L.S. Li, L.W. Ou, G.Q. Chen, and D.W. Jiang, "A sociocultural study of koro epidemics in Guangdong, China," *American Journal of Psychiatry* 145, no. 12 (December 1988): 1538–43.

19   J.M. Roberts, "Belief in the Evil Eye in World Perspective," In *Evil Eye*, ed. C. Maloney (New York: Columbia University Press, 1976), 223–77.

20   D.B. Mumford, "The 'Dhat syndrome': a culturally determined symptom of depression?" *Acta Psychiatrica Scandinavica* 94, no. 3 (September 1996): 163–67.

21   S. Grover, A. Avasthi, S. Gupta, A. Dan, R. Neogi, P.B. Behere, B. Lakdawala, et al., "Phenomenology and beliefs of patients with Dhat syndrome: A nationwide multicentric study," *International Journal of Social Psychiatry* 62, no. 1 (February 2016): 57–66.

22   G.N. Dangerfield, "The symptoms, pathology, causes, and treatment of spermatorrhoea," *The Lancet* 41, no. 1055 (1843): 210–16.

23   P.K. Keel and K.L. Klump, "Are eating disorders culture-bound syndromes? Implications for conceptualizing their etiology," *Psychological Bulletin* 129, no. 5 (September 2003): 747–69.

Chapter 10

1   R. Greenwood, A. Bhalla, A. Gordon, and J. Roberts, "Behaviour disturbances during recovery from herpes simplex encephalitis," *Journal of Neurology, Neurosurgery, and Psychiatry* 46, no. 9 (September 1983): 809–17.

2   E.K. Warrington and T. Shallice, "Category specific semantic impairments," *Brain* 107, pt. 3 (September 1984): 829–54.

3   E.C. Shuttleworth Jr., V. Syring, and N. Allen, "Further observations on the nature of prosopagnosia," *Brain and Cognition* 1, no.3 (July 1982): 307–22.

4   J. Zihl, D. von Cramon, and N. Mai, "Selective disturbance of movement vision after bilateral brain damage," *Brain* 106, pt. 2 (June 1983): 313–40.

5   K. Koch, J. McLean, R. Segev, M.A. Freed, M.J. Berry II, V. Balasubramanian, and P. Sterling, "How much the eye tells the brain," *Current Biology* 16, no. 14 (July 2006): 1428–34.

6   I. Gauthier, P. Skudlarski, J.C. Gore, and A.W. Anderson, "Expertise for cars and birds recruits brain areas involved in face recognition," *Nature Neuroscience* 3, no. 2 (February 2000): 191–97.

7   G.M. Davidson, "A syndrome of time-agnosia," *Journal of Nervous and Mental Disease* 94 (1941): 336–43.

8   S. Thorudottir, H.M. Sigurdardottir, G.E. Rice, S.J. Kerry, R.J. Robotham, A.P. Leff, and R. Starrfelt, "The architect who lost the ability to imagine: the cerebral basis of visual imagery," *Brain Sciences* 10, no. 2 (January 2020): 59.

9   F. Galton, "Statistics of mental imagery," *Mind* 19, no. 1 (July 1880): 301–318.

10 W.F. Brewer and M. Schommer-Aikins, "Scientists are not deficient in mental imagery: Galton revised," *Review of General Psychology* 10, no. 2 (2006): 130–46.

11 A.Z. Zeman, S. Della Sala, L.A. Torrens, V.E. Gountouna, D.J. McGonigle, and R.H. Logie, "Loss of imagery phenomenology with intact visuo-spatial task performance: a case of 'blind imagination,'" *Neuropsychologia* 48, no. 1 (January 2010): 145–55.

12 A. Zeman, M. Dewar, and S. Della Sala, "Lives without imagery—Congenital aphantasia," *Cortex* 73 (December 2015): 378–80.

13 B. Faw, "Conflicting intuitions may be based on differing abilities evidence from mental imaging research," *Journal of Consciousness Studies* 16, no. 2 (2009): 45–68.

14 G. Ganis, W.L. Thompson, and S.M. Kosslyn, "Brain areas underlying visual mental imagery and visual perception: an fMRI study," *Brain Research: Cognitive Brain Research* 20, no. 2 (July 2004): 226–41.

15 A.Z. Zeman, S. Della Sala, L.A. Torrens, V.W. Gountouna, D.J. McGonigle, and R.H. Logie, "Loss of imagery phenomenology," 145-55.

16 E.C. Shuttleworth Jr, V. Syring, and N. Allen, "Further observations on the nature of prosopagnosia," *Brain and Cognition* 1, no. 3 (July 1982): 307–22.

Chapter 11

1   M. Murdoch, J. Hill, and M. Barber, "Strangled by Dr Strangelove? Anarchic hand following a posterior cerebral artery territory ischemic stroke," *Age and Ageing* 50, no. 1 (January 2021): 263–64.

2   L.A. Scepkowski and A. Cronin-Golomb, "The alien hand: cases, categorizations, and anatomical correlates," *Behavioral and Cognitive Neuroscience Reviews* 2, no. 4 (December 2003): 261–77.

3   M. Ali, K. VandenBerg, L.J. Williams, L.R. Williams, M. Abo, F. Becker, A. Bowen, et al., "Predictors of poststroke aphasia recovery: a sytematic review-informed individual participant data meta-analysis," *Stroke* 52, no. 5 (May 2021): 1778–87.

4   C. Ochipa, L.J. Rothi, and K.M. Heilman, "Ideational apraxia: a deficit in tool selection and use," *Annals of Neurology* 25, no. 2 (February 1989): 190–93.

5   K. Poeck, "Ideational apraxia," *Journal of Neurology* 230, no. 1 (1983): 1-5.

6   A. Dressing, C.P. Kaller, M. Martin, K. Nitschke, D. Kuemmerer, L.A. Beume, C.S.M. Schmidt, et al., "Anatomical correlates of recovery in apraxia: a longitudinal lesion-mapping study in stroke patients," *Cortex* 142 (September 2021): 104–21.

7   R.G. Gross and M. Grossman, "Update on apraxia," *Current Neurology and Neuroscience Reports* 8, no. 6 (2008): 490–96.

8   M. Kinsbourne and E.K. Warrington, "A study of finger agnosia," *Brain* 85 (March 1962): 47–66.

9   E. Rusconi and R. Cubelli, "The making of a syndrome: the English translation of Gerstmann's first report," *Cortex* 117 (August 2019): 277–83.

10 J. Gerstmann, "Syndrome of finger agnosia, disorientation for right and left, agraphia and acalculia," *Archives of Neurology & Psychiatry* 44, no. 2 (1940): 398–408.

11 E. Rusconi, P. Pinel, E. Eger, D. LeBihan, B. Thirion, S. Dehaene, and A. Kleinschmidt, "A disconnection account of Gerstmann syndrome: functional neuroanatomy evidence," *Annals of Neurology* 66, no. 5 (November 2009): 654–62.

Chapter 12

1   M. Vilela, D. Fernandes, T. Salazar Sr., C. Maio, and A. Duarte, "When Alice took sertraline: a case of sertraline-induced Alice in Wonderland syndrome," *Cureus* 12, no. 8 (August 2020): e10140.

2   J.D. Blom, "Alice in Wonderland syndrome: a systematic review," *Neurology Clinical Practice* 6, no. 3 (June 2016): 259–70.

3   G. Mastria, V. Mancini, A. Viganò, and V. Di Piero, "Alice in Wonderland syndrome: a clinical and pathophysiological review," *Biomed Research International* (December 2016): 1–10.

4   T.W. Abell, "Remarkable case of illusive vision," *The Boston Medical and Surgical Journal* 33 (1845): 409–13.

5   M.E. McNamara, R.C. Heros, and F. Boller, "Visual hallucinations in blindness: the Charles Bonnet syndrome," *International Journal of Neuroscience* 17, no. 1 (July 1982): 13–15.

6   R.K. Siegel, "Hostage hallucinations. Visual imagery induced by isolation and life-threatening stress," *The Journal of Nervous and Mental Disease* 172, no. 5 (May 1984): 264–72.

7   O.J. Mason and F. Brady, "The psychotomimetic effects of short-term sensory deprivation," *The Journal of Nervous and Mental Disease* 197, no. 10 (October 2009): 783–85.

8   L.B. Merabet, D. Maguire, A. Warde, K. Alterescu, R. Stickgold, and A. Pascual-Leone, "Visual hallucinations during prolonged blindfolding in sighted subjects," *Journal of Neuro-Ophthalmology* 24, no. 2 (June 2004): 109–13.

9   D.H. Ffytche, R.J. Howard, M.J. Brammer, A. David, P. Woodruff, and S. Williams, "The anatomy of conscious vision: an fMRI study of visual hallucinations," *Nature Neuroscience* 1, no. 8 (December 1998): 738–42.

10 C.E. Sluzki, "Saudades at the edge of the self and the merits of 'portable families,'" *Transcultural Psychiatry* 45, no. 3 (September 2008): 379–90.

11 "Two in Five Americans Say Ghosts Exist—and One in Five Say They've Encountered One," *YouGovAmerica*, last modified October 21, 2021, https://today.yougov.com/topics/entertainment/articles-reports/2021/10/21/americans-say-ghosts-exist-seen-a-ghost.

12 K.S. Kamp and H. Due, "How many bereaved people hallucinate about their loved one? A systematic review and meta-analysis of bereavement hallucinations," *Journal of Affective Disorders* 243 (January 2019): 463–76.

13 P.R. Olson, J.A. Suddeth, P.J. Peterson, and C. Egelhoff, "Hallucinations of widowhood," *Journal of the American Geriatrics Society* 33, no. 8 (1985): 543–47.

14 K. Koch, J. McLean, R. Segev, M.A. Freed, M.J. Berry II, V. Balasubramanian, and P. Sterling, "How much the eye tells the brain," *Current Biology* 16, no. 14 (July 2006): 1428–34.

INDEX

ABOUT THE AUTHOR

Marc Dingman received his PhD in Neuroscience in 2013 from The Pennsylvania State University. Since then, he has been a faculty member in the Biobehavioral Health Department at Penn State, where he teaches courses in neuroscience and the health sciences. He also spends much of his free time teaching people about neuroscience through his website (www.neurochallenged.com) and his popular YouTube series, *2-Minute Neuroscience*. He lives outside State College, Pennsylvania, with his wife and two children.

Learn more from Marc's website and his YouTube video series, *2-Minute Neuroscience*

If you want to learn more about the brain, my website (www .neurochallenged.com) is a good place to start. It has links to all my *2-Minute Neuroscience* videos, along with hundreds of articles and a 500-plus-word glossary. And, if you have questions you can't find an answer to there, feel free to email me at neurochallenged@gmail.com.